ACINETOBACTER

Microbiology, Epidemiology, Infections, Management

Edited by

E. Bergogne-Bérézin
Department of Microbiology
Bichat-Claude Bernard Hospital
Paris, France

M.L. Joly-Guillou
Department of Microbiology
Bichat-Claude Bernard Hospital
Paris, France

K.J. Towner
Department of Microbiology and PHLS Laboratory
University Hospital
Queen's Medical Center
Nottingham, United Kingdom

CRC Press
Boca Raton New York London Tokyo

Library of Congress Cataloging-in-Publication Data

Acinetobacter: microbiology, epidemiology, infections, management/ edited by E. Bergogne-Bérézin, M.-L. Joly-Guillou, K.J. Towner.
p. cm.
Includes bibliographical references and index.
ISBN 0-8493-9223-3 (permanent paper)
1.Acinetobacter infections. 2. Nosocomial infections. I. Bergogne-Bérézin, E. II. Joly-Guillou, M.-L. III. Towner, K.J.
QR201.A16A25 1995Z
616′.0145—dc20

95-30812
CIP

This book contains information obtained from authentic and highly regarded sources. Reprinted material is quoted with permission, and sources are indicated. A wide variety of references are listed. Reasonable efforts have been made to publish reliable data and information, but the author and the publisher cannot assume responsibility for the validity of all materials or for the consequences of their use.

Direct all inquiries to CRC Press, Inc., 2000 Corporate Blvd., N.W., Boca Raton, Florida 33431.

International Standard Book Number 0-8493-9223-3
Library of Congress Card Number 95-30812
Printed in the United States of America 2 3 4 5 6 7 8 9 0
Printed on acid-free paper

PREFACE

The concept of the genus *Acinetobacter* was first proposed by microbiologists working in France in 1954. Although known to be common in the environment, these bacteria attracted little attention from clinicians for over 20 years. Occasional reports of *Acinetobacter* infections occurred as curiosities in the medical literature, but in general, these bacteria were considered to be saprophytes of little clinical importance. This situation changed during the 1970s following the introduction of powerful new antibiotics and the increasing use of invasive procedures (catheters, central lines, exploratory techniques, etc.) in hospital intensive care units to treat severely debilitated patients. This period coincided with increasing awareness of the pathogenic role of *Acinetobacter* spp., particularly in nosocomial infection, and the propensity of this genus to develop resistance rapidly to even the most potent antibiotics. This second period has lasted from the mid-1970s until today, during which time new patterns of resistance have emerged in this genus and have become disseminated in numerous hospitals. Many clinicians still lack an appreciation of the potential importance of these organisms in hospitals, from both the patient and the economic viewpoint. In this book we have tried to gather together the sum total of knowledge, some of which is previously unpublished, regarding *Acinetobacter* spp., their microbiology, epidemiology, infections, resistance to antibiotics, and potential therapeutic approaches. All clinicians and microbiologists should assimilate the lessons that *Acinetobacter* spp. are teaching us, for if we are unable ultimately to control *Acinetobacter* spp., the overall struggle against resistant nosocomial bacteria is in danger of being lost.

E. Bergogne-Bérézin
Paris, France
M.-L. Joly-Guillou
Paris, France
K.J. Towner
Nottingham, United Kingdom

The Editors

Professor E. Bergogne-Bérézin, M.D., Ph.D., has been Professor of Microbiology at the Medical School Xavier-Bichat (University of Paris VII) since 1967 and is also Head of the Department of Microbiology at the Bichat-Claude Bernard Hospital in Paris. She is a consultant in pharmacology and toxicology for the French Ministry of Health, and is also a member of the Advisory Boards of three multinational pharmaceutical companies. Professor Bergogne-Bérézin has published well over 100 original scientific articles and has written or contributed to numerous books and review articles. She is a member of the Editorial Advisory Boards of the *Journal of Chemotherapy,* the *International Journal of Antimicrobial Agents, Microbiology (Europe),* and *La Presse Médicale.*

Professor Bergogne-Bérézin's current research interests include *in vitro* studies of antibiotics (including their pharmacokinetic behaviour and mechanisms of antibiotic resistance in bacteria), investigations of the behaviour and pathogenesis of infection of nosocomial pathogens (particularly *Acinetobacter* spp.), animal models of pneumonia, and studies of intestinal microflora and intestinal bacterial adhesion.

M.-L. Joly-Guillou, M.D., Ph.D., has been Associate Professor of Microbiology at the Medical School Xavier-Bichat (University of Paris VII) since 1990. Dr. Joly-Guillou is a member of the French Society of Microbiology, the Society of Infectious Disease, and the Association of Pasteur Institute for former pupils.

Dr. Joly-Guillou's current research interests include *in vitro* studies of antibiotics and resistance mechanisms, epidemiology and clinical studies, virulence and development of experimental infection in relation with INSERM U13 (animal models of pneumonia).

Kevin J. Towner, Ph.D., graduated from the University of Reading, United Kingdom, in 1974 with a degree in Microbiology. He then spent 3 years in London completing a Ph.D. doctoral thesis on the genetics, antibiotic resistance, and molecular biology of *Acinetobacter* spp., followed by a move to the Department of Microbiology and PHLS Laboratory at University Hospital in Nottingham. Dr. Towner currently holds the post of Consultant Clinical Scientist and Head of the Molecular Biology Section within the joint academic and diagnostic microbiology laboratories in Nottingham. Dr. Towner has published over 70 original scientific papers in refereed journals, and has either written, edited, or contributed to numerous books and review articles. He has served as an Editor of the *Journal of Antimicrobial Chemotherapy,* and is currently an Editor of the *Journal of Medical Microbiology* and a member of the Editorial Advisory Board for *Clinical Microbiology and Infection.*

Dr. Towner's current research interests include the molecular aspects of antimicrobial resistance in bacteria, the application of new molecular techniques to the identification, typing, and epidemiology of microorganisms (particularly those of nosocomial importance), and all aspects of the biology and clinical importance of *Acinetobacter* spp.

Contributors

Professor S. G. B. Amyes
Department of Medical Microbiology
University of Edinburgh Medical School
Edinburgh, Scotland

Professor E. Bergogne-Bérézin
Department of Microbiology
Bichat-Claude Bernard Hospital
Paris, France

Professor C. Brun-Buisson
Medical Intensive Care Unit
Henri-Mondor Hospital
Creteil, France

Professor Jean Chastre
Medical Intensive Care Unit
Bichat-Claude Bernard Hospital
Paris, France

Dr. Lenie Dijkshoorn
Department of Medical Microbiology
Academisch Ziekenhuis Leiden
Leiden, The Netherlands

Dr. M.-L. Joly-Guillou
Department of Microbiology
Bichat-Claude Bernard Hospital
Paris, France

Dr. Itzchak Levi
Chaim Sheba Medical Center
Tel-Aviv University
Tel Hashomer, Israel

Professor Ethan Rubinstein
Chaim Sheba Medical Center
Tel-Aviv University
Tel Hashomer, Israel

Dr. Kevin J. Towner
Department of Microbiology and PHLS Laboratory
University Hospital
Queen's Medical Centre
Nottingham, United Kingdom

Dr. Jean-Louis Trouillet
Medical Intensive Care Unit
Bichat-Claude Bernard Hospital
Paris, France

Dr. Albert Vuagnat
Medical Intensive Care Unit
Bichat-Claude Bernard Hospital
Paris, France

Dr. H.-K. Young
Department of Biological Sciences
University of Dundee
Dundee, Scotland

Dr. Michel Wolff
Infectious Diseases Intensive Care Unit
Bichat-Claude Bernard Hospital
Paris, France

CONTENTS

ACINETOBACTER

Microbiology, Epidemiology, Infections, Management

Chapter 1

Introduction – History and Importance of *Acinetobacter* spp., Role in Infections, Treatment and Cost Implications

E. Bergogne-Bérézin, M.-L. Joly-Guillou, and K.J. Towner

CONTENTS

0-8493-9223-3/95/$0.00+$.50

1.1 INTRODUCTION

There is increasing interest in developed countries in the control of hospital-acquired infection. Amongst nosocomial pathogens, multiply resistant Gram-negative bacilli have proved to be a particular problem over the last 30 years. The therapeutic introduction of early antibiotics, mainly to control Gram-positive infections, was followed by the emergence of Gram-negative bacteria in the late 1960s, and an increasing incidence during the 1970s of resistant Enterobacteriaceae involved in nosocomial infections. As newer broad-spectrum antibiotics were introduced into hospitals, strictly aerobic Gram-negative bacilli became more important, particularly *Pseudomonas aeruginosa, Xanthomonas (Stenotrophomonas) maltophilia* and *Acinetobacter* spp. These 'newer' pathogens possess several characteristics that enhance their ability to acquire and express antibiotic resistance: they are non-fastidious saprophytic organisms that do not require unusual nutrients and substrates; they survive in poor conditions, in dry environments, and even in the presence of antiseptics; they offer a permeability barrier which renders them intrinsically resistant to some antibiotics; and they possess a great diversity of plasmids and chromosomally mediated enzymes that give them the potential to become multiply resistant organisms. In addition, they have many opportunities for transmission between patients, either via human reservoirs or via inanimate materials in the hospital environment. This initial broad profile defines the important characteristics of *Acinetobacter* spp. which form the subject of this book.

1.2 HISTORY AND TAXONOMIC BACKGROUND

Before a particular bacterial species or genus can be implicated in infections and recognised as potentially pathogenic in man, it must be clearly defined, classified and identified. The genus *Acinetobacter* has undergone extensive and confusing taxonomic changes over many years, and these organisms have been classified previously in various genera (see Appendix I), the most common being designated *Bacterium anitratum* (Schaub et al., 1948), *Herellea vaginicola* and *Mima polymorpha* (Debord, 1939), *Achromobacter, Alcaligenes, Neisseria, Micrococcus calcoaceticus, Diplococcus,* 'B5W' and *Cytophaga* (Juni, 1972). The genus *Acinetobacter,* as conceived originally by Brisou and Prévôt (1954), included oxidase-positive (*Moraxella*) and oxidase-negative strains (Piéchaud, 1961; Bøvre and Henriksen, 1976). In 1961, in the *Annales de l'Institut Pasteur,* Piéchaud described strictly aerobic, oxidase-negative, Gram-negative coccobacilli, that he considered should be classified as either *Moraxella glucidolytica* (including 'B5W' and *Bacterium anitratum*), for strains acidifying glucose, or *Moraxella lwoffii*. Both species were penicillin resistant, and the identification of strains was based on phenotypic characters (cultural and metabolic) and morphology, which was described in precise detail, including cell division stages revealed by the Piekarski-Robinow dyeing technique. Piéchaud stated at the end of the article "Rien ne justifie l'inclusion de *M. lwoffii*, de *M. glucidolytica* et de leurs variétés dans un genre *Acinetobacter* (Brisou, 1957), groupe hétérogène (dont la plupart des espèces sont décrites comme anaérobies facultatives) tels *Acinetobacter* stenohalis, butyri, eurydice, delmarvae, marshalli, metalcaligenes...."*

It was only in 1971 that the Subcommittee on the Taxonomy of *Moraxella* and Allied Bacteria proposed that the genus *Acinetobacter* should include only the oxidase-negative strains. In *Bergey's Manual of Systematic Bacteriology* (Juni, 1984), the genus *Acinetobacter* comprised one species, *A. calcoaceticus*, and two varieties: var. *anitratus* (formerly *Herella vaginicola*) and var. *lwoffii* (formerly *Mima polymorpha*), grouped in the family Neisseriaceae. This erratic and changing taxonomy hindered progress in understanding the pathogenic role of *Acinetobacter* spp. in humans (Juni, 1984; Bergogne-Bérézin et al., 1987), but for more than a decade, the latter classification has been of great help in studying the epidemiology of *Acinetobacter* spp. and identifying the pathogenic role of acinetobacters in infections, and resulted in recognition of the fact that *A. calcoaceticus* var. *anitratus* was involved more frequently in infections and was the most resistant variety of *Acinetobacter* spp. (Glew et al., 1977; Vieu et al., 1979; Godineau-Gauthey et al., 1988; Bergogne-Bérézin and Joly-Guillou, 1991). More recently, modern methods of taxonomy (genetic transformation, DNA hybridisation and rRNA sequence comparisons) have resulted in a new classification of *Acinetobacter* spp., which comprises 19 DNA-DNA homology groups (genomic species), of which, to date, 7 have been given species names (see Chapter 2).

1.3 BIOLOGY OF *Acinetobacter* spp.

Strains of *Acinetobacter* are ubiquitous non-fermentative organisms that are distributed widely in nature (see Chapter 2). In appearance, acinetobacters have the form of short, plump, Gram-negative rods, typically 1.0–1.5 μm × 1.5–2.5 μm in the logarithmic phase of growth, but often becoming more coccoid in the stationary phase (Figures 1.1 and 1.2). They are non-fastidious organisms that are easily grown and can utilise a wide variety of substrates as sole carbon sources. All

*"Nothing justifies the inclusion of *M. lwoffii*, *M. glucidolytica* and their varieties in a genus *Acinetobacter* (Brisou, 1957), an heterogeneous group (including most species described as facultatively anaerobic) such as *Acinetobacter*...."

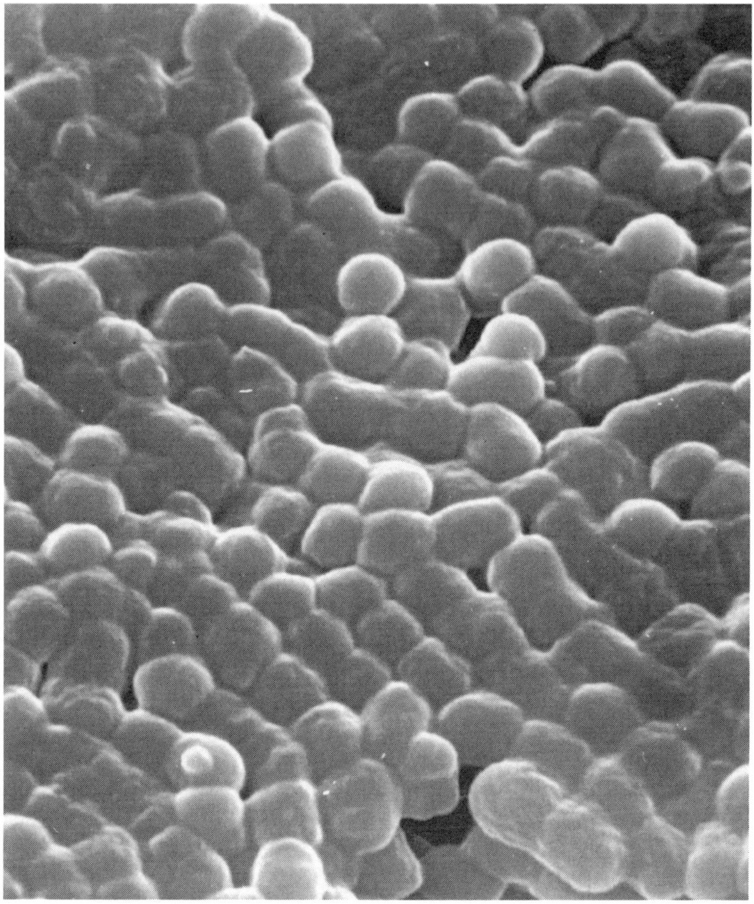

FIGURE 1.1
Scanning electron micrograph of *A. baumannii* type strain ATCC 19606 (final magnification of 18,000). Prepared and photographed by A. Shelton.

members of the genus are strict aerobes and can grow at a wide range of temperatures, forming smooth, sometimes mucoid colonies on solid media (Figure 1.3). Clinical isolates will grow at 37°C, but some environmental isolates prefer incubation temperatures of 20-30°C. All strains of *Acinetobacter* are oxidase-negative and catalase-positive, and it is the negative oxidase reaction which serves to distinguish the genus from other genera grouped in the family Neisseriaceae. The identification, ecology and characteristics of acinetobacters are discussed in more detail in Chapters 2 and 3, but it is apparent that although the genus, taken as a whole, is distributed widely, there is a significant population difference between the acinetobacters found in clinical and other

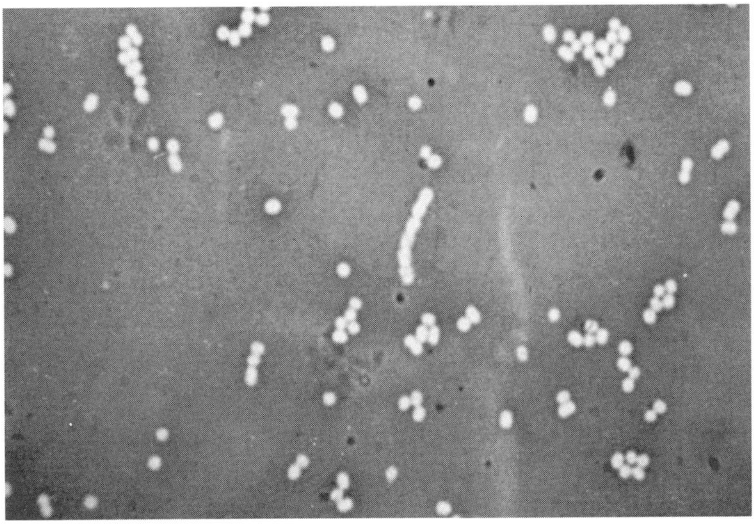

FIGURE 1.2
Negatively stained wet mount of *A. baumannii* type strain ATCC 19606.

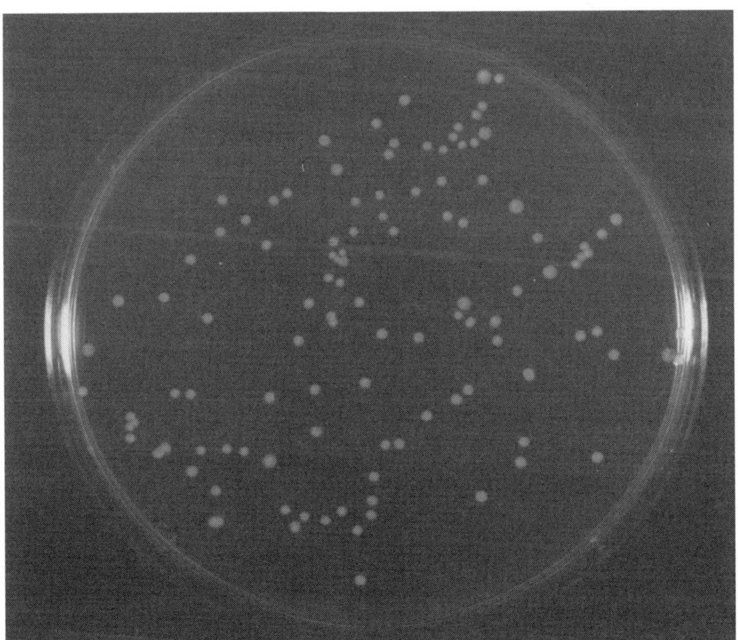

FIGURE 1.3
Colonies of *A. baumannii* type strain ATCC 19606 growing on a plate of nutrient agar following incubation for 24 h at 37°C.

environments. Each population appears to be characterised by predominant groups of genomic species; thus, for example, spp. 2 (*A. baumannii*) and 3 are found predominantly in clinical environments, while spp. 7 (*A. johnsonii*) and 8/9 (*A. lwoffii*) are found predominantly in non-clinical environments. All three of the major modes of chromosomal gene transfer have been demonstrated in *Acinetobacter* spp. (see Chapter 2), and genetic interchange between populations is clearly possible, a fact that may have implications for the spread of antibiotic resistance genes.

1.4 THE ROLE OF *Acinetobacter* spp. IN NOSOCOMIAL INFECTION

It is now recognised that *Acinetobacter* strains play a significant role in colonisation and infection of patients admitted to hospital intensive care units (ICUs). They may be recovered from the hospital environment, from colonised or infected patients, and from the hands of staff (see Chapter 4). As opportunistic pathogens, they are responsible for nosocomial infections, including bacteraemia, urinary tract infection and secondary meningitis (see Chapter 5), but their predominant role is as agents of nosocomial pneumonia, particularly ventilator-associated pneumonia (see Chapter 6). New molecular typing methods have now been developed that allow the recognition of genotypically related strains involved in outbreaks of hospital infection (see Chapters 3 and 4).

One of the main reasons for increased interest in this genus is the fact that these bacteria are often responsible for difficult-to-treat infections because of their frequent multiple resistance to the major antibiotics available for the treatment of nosocomial infection. Various mechanisms of resistance have been recognised in these bacteria (see Chapter 8), and combination therapy is usually required for effective treatment of *Acinetobacter* nosocomial infections (Bergogne-Bérézin and Joly-Guillou, 1991; see Chapter 7). Risk factors for acquisition of multiresistant acinetobacters include the four main groups of factors inherent to hospitals and patients in ICUs:

1. Environmental factors, such as the presence of environmental sources, the use of increasingly complicated invasive procedures, and the admission of patients to areas of special concern, such as ICUs, abdominal or thoracic surgery units and haemodialysis units
2. Patient-associated risk factors, including debilitation with impaired immune defense mechanisms and prolonged hospitalisation with increased exposure to a contaminated environment
3. Various therapeutic factors which may contribute to the impairment of a patient's resistance, including corticosteroids, anti-metabolites, anti-cancer chemotherapy, and radiation therapy

4. The use of broad-spectrum antibiotics, which are now regarded as having played a major role in selecting strains of *Acinetobacter* with the most impressive patterns of antibiotic resistance. These factors are examined and analysed in detail in subsequent chapters.

1.5 CLINICAL IMPORTANCE OF *Acinetobacter* spp.

Acinetobacter has been isolated from various types of opportunistic infection, including septicaemia, pneumonia, endocarditis, meningitis, skin and wound sepsis, and urinary tract infection (French et al., 1980; Bergogne-Bérézin et al., 1987; Ng et al., 1989; Joly-Guillou and Bergogne-Bérézin, 1992; see Chapter 5). Although the organism is associated predominantly with nosocomial infection, some cases of community-acquired infections have been reported, indicative of the primary pathogenicity of these organisms (Rudin et al., 1979; Hoffmann et al., 1982; Anstey et al., 1992). The distribution by site of *Acinetobacter* infection does not differ from other nosocomial Gram-negative bacteria, with the main sites of infection in several surveys (e.g., Glew et al., 1977; Joly-Guillou and Bergogne-Bérézin, 1992; Joly-Guillou et al., 1992) being the lower respiratory tract and the urinary tract, with rates of infection ranging from 15–30% of total infections caused by acinetobacters (clinical manifestations of *Acinetobacter* infection are described in detail in Chapters 5 and 6).

Various risk factors predisposing to severe infection have been identified, some of which are also common to organisms causing other nosocomial infections. Susceptible patients include those with severe underlying disease (e.g., malignancy, burns, immunosuppression and major surgery), particularly the elderly, although outbreaks of *Acinetobacter* infection have also been described in neonates (Stone and Das, 1986; Ng et al., 1989). The use of antibiotics probably alters the normal flora and results in the selection of resistant microorganisms such as *Acinetobacter* spp. In hospitals in France, *Acinetobacter* spp. now account for 9.7% of nosocomial infections (Joly-Guillou et al., 1992).

Although considered originally to be relatively low-grade pathogens (Pedersen et al., 1970; Juni, 1978; Smego, 1985), *Acinetobacter* spp. may assume greater pathogenic importance under certain conditions. Several characteristics of these bacteria seem to play a role in enhancing the virulence of strains involved in infections: (1) the presence of a polysaccharide capsule formed of L-rhamnose and D-glucose (Juni, 1984), which renders the surface of strains more hydrophilic (although hydrophobicity is higher in *A. baumannii* strains isolated from catheters or tracheal devices); (2) the property of adhesion to human epithelial cells in the presence of fimbriae, which correlates with twitching motility, and/or capsular polysaccharide; (3) the production of certain enzymes (butyrate esterase, caprylate esterase and leucine arylamidase), which seem to be involved in damaging tissue lipids; and (4) the potentially toxic role of the lipopolysaccharide component of the cell wall and the presence of lipid A (Avril and Mesnard, 1991). Strains of *A. baumannii* have limited virulence in mice (LD_{50} values of 10^6–10^8 cfu/mouse) when inoculated intraperitoneally, even in neutropenic mice. However, recent experimental studies in a mouse model of *Acinetobacter* pneumonia have shown that this pneumonia closely resembles that which occurs in humans (Joly-Guillou et al., 1994).

1.6 ANTIBIOTIC RESISTANCE IN *Acinetobacter* spp.

One of the most striking features of *Acinetobacter* spp. since their recognition as significant human pathogens has been their apparent ability to develop antibiotic resistance extremely rapidly in response to challenge with new antibiotics (see Chapters 7 and 8). It is probable that it is this ability, combined with widespread use of radical broad-spectrum antibiotics in the hospital environment, that is responsible for their increasing importance as nosocomial pathogens. As surgical procedures and mechanical interventions become even more complex, bacteria which have the capability to develop resistance to the radical antibiotics used in conjunction with these treatments will become more predominant. As discussed in Chapters 7 and 8, *Acinetobacter* spp. have already developed resistance to broad-spectrum cephalosporins, 4-quinolones and, to some extent, the carbapenems. Mechanisms of gene transfer for mobilisation of resistance genes between different members of the genus (and possibly to unrelated bacteria of greater pathogenicity) are known to exist (see Chapters 2 and 8), and it is not clear how infections caused by these organisms will be controlled once complete resistance to these three groups of antibiotics has evolved.

1.7 ECONOMICS OF NOSOCOMIAL INFECTION CAUSED BY *Acinetobacter* spp.

The worldwide antimicrobial market ($20.5 million in 1992) ranks second after cardiovascular drugs and is dominated by cephalosporins (44% of all antibiotic sales). Wide differences exist between the prescription of antibiotics in community practice and antibiotic usage in hospitals, with respect to doses, available drugs and concepts in antibiotic therapy. A rough evaluation in France indicates that in 1993, 39% of total antibiotics were prescribed in the community, with 23% of antibiotics used for treating nosocomial infections, and 39% used for antibiotic prophylaxis. At Bichat Hospital in Paris, the three main antibiotic classes used for treating nosocomial infection are aminoglycosides, fluoroquinolones and third-generation cephalosporins. The use of these compounds has considerable implications in terms of the financial problems currently being faced by many hospitals.

Pharmacoeconomics is a developing subject, and studies of the economics of drug therapy are found increasingly in the medical literature (e.g., Davey et al., 1993; 1994). In order to establish cost-effectiveness, the following factors need to be investigated: (1) the direct and indirect costs of treatment, as well as potential benefits (if any); (2) the effectiveness of therapy, based arbitrarily on either survival rates or an improvement in the patient in terms of a change of grade in severity scores, decreased number of days of positive blood cultures, etc.; and (3) the financial savings that can be obtained by an infection control programme.

Modern methods for estimating the costs of nosocomial infection include various components, of which bacteriology and antimicrobial usage normally represent minor parameters. However, additional costs for typing of strains (for epidemiological studies), repeat cultures for studying reoccurrence of infection, control of infection studies, determination of antibiograms and *in vitro* testing of antibiotic combinations, etc., should be added to the direct costs of bacteriological analysis. With opportunistic pathogens such as *Acinetobacter* spp. it can be postulated that these additional costs may be substantial. Similarly, antibiotic costs for treating *Acinetobacter* nosocomial infection can be much higher than would be expected on the basis of antibiotic use only. As discussed in subsequent chapters of this book, antibiotic therapy for *Acinetobacter* infections includes new and expensive drugs, as well as the frequent changes of antibiotic combinations that are required when emergence of resistance occurs during therapy. To the direct antibiotic costs should be added other well-known factors such as prolongation of stay (estimated at an average of 13.3 days for nosocomial infections) and various extra costs attributable to infection, such as the costs of cross-infection resulting from the spread and persistence in the hospital ecosystem of these highly resistant bacteria. Thus, in economic terms, infections caused by *Acinetobacter* spp. should be regarded as 'high-cost infections'. Outbreaks of nosocomial infection caused by such multiresistant organisms represent a substantial increased cost of hospital care that is additional to the 'normal' costs of a patient's care. Such a 'marginal' indirect cost is unpredictable, and is hardly ever identified, but it forms a significant component of the financial problems currently being confronted by many hospitals.

1.8 SUMMARY

For many years, members of the genus *Acinetobacter* received relatively little attention from clinical microbiologists, partly because of their confused taxonomic history and partly because of their perceived role as saprophytic bacteria that caused only very occasional opportunistic infections in severely debilitated patients. The main factors responsible for the increased clinical significance now attached to these organisms seem to have been the therapeutic introduction of new potent broad-spectrum antimicrobials during the 1970s and 1980s, coupled with the increasing use of complex invasive procedures in patients confined to ICUs. Members of the genus *Acinetobacter* seem to have a remarkable ability to develop resistance to even the most potent antimicrobials, and have rapidly filled the ecological niche left vacant by the elimination of competing bacteria with broad-spectrum compounds. The absence of competing bacteria has also enhanced the ability of acinetobacters to colonise and subsequently cause a wide range of nosocomial infections in patients, some of which may be life-threatening, and most of which are difficult and expensive to treat. The succeeding chapters in this book bring together the various relevant facets of *Acinetobacter* in an attempt to provide as complete an account as possible of the significant clinical problems now posed by these bacteria, as well as the possible therapeutic approaches that might be used to address them.

REFERENCES

Anstey, M.N., Currie, B.J. and Withrall, K.M. (1992). Community-acquired *Acinetobacter* pneumonia in the Northern territory of Australia. *Clin. Infect. Dis.* 14, 83–91.

Avril, J.L. and Mesnard, R. (1991). Factors influencing the virulence of *Acinetobacter*. In *The Biology of* Acinetobacter, pp. 77–82. K.J. Towner, E. Bergogne-Bérézin and C.A. Fewson (Eds). Plenum Press, New York.

Bergogne-Bérézin, E. and Joly-Guillou, M.-L. (1991). Antibiotic resistance mechanisms in *Acinetobacter*. In *The Biology of* Acinetobacter, pp. 83–115. K.J. Towner, E. Bergogne-Bérézin and C.A. Fewson (Eds). Plenum Press, New York.

Bergogne-Bérézin, E., Joly-Guillou, M.-L. and Vieu, J.F. (1987). Epidemiology of nosocomial infections due to *Acinetobacter calcoaceticus*. *J. Hosp. Infect.* 10, 105–113.

Bøvre, K. and Henriksen, S.D. (1976). Minimal standards for description of new taxa within the genera *Moraxella* and *Acinetobacter*: proposal by the Subcommittee on *Moraxella* and Allied Bacteria. *Int. J. Syst. Bacteriol.* 26, 92–96.

Brisou, J. and Prévôt, A.R. (1954). Etudes de systématique bactérienne. X-Révision des espèces réunies dans le genre *Achromobacter*. *Ann. Inst. Pasteur* 86, 722–728.

Davey, P.G., Parker, S.E. and Malek, M.M. (1993). Pharmacoeconomics of antimicrobial prophylaxis. *J. Antimicrob. Chemother.* 31 (Suppl. B), 107–118.

Davey, P.G., Malek, M., McMurdo, M. and Tarnow-Mordi, W. (1994). Economics of infection at the extremes of age. *J. Antimicrob. Chemother.* 34 (Suppl. A), 129–143.

Debord, G.G. (1939). Organisms invalidating the diagnosis of gonorrhea by the smear method. *J. Bacteriol.* 38, 119–120.

French, G.L., Casewell, M.W., Roncoroni, A.J., Knight, S. and Phillips, I. (1980). A hospital outbreak of antibiotic-resistant *Acinetobacter anitratus*: epidemiology and control. *J. Hosp. Infect.* 1, 125–131.

Glew, R.H., Moellering, R.C. and Kunz, L.J. (1977). Infections with *Acinetobacter calcoaceticus* (*Herellea vaginicola*). Clinical and laboratory studies. *Medicine* 56, 79–87.

Godineau-Gauthey, N., Lesage, D., Tessier, F., Kollia, D. and Daguet, G.L. (1988). *Acinetobacter calcoaceticus* variété *anitratus* ou *Acinetobacter baumannii*. Etude de la sensibilité aux antibiotiques chez 65 souches hospitalières. *Méd. Mal. Infect.* ii, 124.

Hoffmann, S., Mabeck, C.E. and Vejlsgaard, R. (1982). Bacteriuria caused by *Acinetobacter calcoaceticus* biovars in a normal population and in general practice. *J. Clin. Microbiol.* 16, 443–451.

Joly-Guillou, M.-L. and Bergogne-Bérézin, E. (1992). In vitro activity of sparfloxacin, pefloxacin, ciprofloxacin and temafloxacin against clinical isolates of *Acinetobacter* spp. *J. Antimicrob. Chemother.* 29, 466–468.

Joly-Guillou, M.-L, Decré, D., Wolff, M. and Bergogne-Bérézin, E. (1992). A*cinetobacter* spp: clinical epidemiology in 89 intensive care units. A retrospective study in France during 1991. In *Abstr. 2nd Int. Conf. Prevention of Infection* (Nice, France).

Joly-Guillou, M.-L., Wolff, M., Walker F. and Vallée, E. (1994). A mouse model of *Acinetobacter baumannii* pneumonia. In *Abstr. 3rd Int. Symp. Biol.* Acinetobacter (Edinburgh).

Juni, E. (1972). Interspecies transformation of *Acinetobacter*: genetic evidence for a ubiquitous genus. *J. Bacteriol.* 112, 917–931.

Juni, E. (1978). Genetics and physiology of *Acinetobacter*. *Annu. Rev. Microbiol.* 32, 349–371.

Juni, E. (1984). Genus III. *Acinetobacter* Brisou et Prévot 1954. In *Bergey's Manual of Systematic Bacteriology*, Vol. 1, pp. 307–308. N.R. Krieg and J.G. Holt (Eds). Williams & Wilkins, Baltimore.

Ng, P.C., Herrington, R.A., Beane, C.A., Ghoneim, A.T. and Dear P.R. (1989). An outbreak of *Acinetobacter* septicaemia in a neonatal intensive care unit. *J. Hosp. Infect.* 14, 363–368.

Pedersen, M.M., Marso, E. and Pickett, M.J. (1970). Non-fermentative bacilli associated with man. III. Pathogenicity and antibiotic susceptibility. *Am. J. Clin. Pathol.* 54, 178–192.

Piéchaud, M. (1961). Le groupe Moraxella. A propos de B5W-*Bacterium anitratum*. *Ann. Inst. Pasteur* 100, 74–85.

Rudin, M.L., Michael, J.R. and Huxley E.J. (1979). Community-acquired *Acinetobacter* pneumonia. *Am. J. Med.* 67, 39–43.

Schaub, I.G. and Hauber, F.D. (1948). A biochemical and serological study of a group of identical unidentifiable Gram-negative bacilli in human sources. *J. Bacteriol.* 56, 379–385.
Smego, R.A. (1985). Endemic nosocomial *Acinetobacter calcoaceticus* bacteremia. Clinical significance, treatment and prognosis. *Arch. Int. Med.* 145, 2174–2179.
Stone, J.M. and Das, B.C. (1986). Investigation of an outbreak of infection with *Acinetobacter calcoaceticus* in a special care baby unit. *J. Hosp. Infect.* 7, 42–48.
Vieu, J.F., Minck, R. and Bergogne-Bérézin, E. (1979). Bactériophages et lysotypie d'Acinetobacter. *Ann. Microbiol. Inst. Pasteur* 130A, 405–406.

Chapter 2

BIOLOGY OF *Acinetobacter* spp.

K.J. Towner

CONTENTS

0-8493-9223-3/95/$0.00+$.50

2.1 INTRODUCTION

Gram-negative non-fermentative bacteria now recognised as belonging to the genus *Acinetobacter* have been classified previously under a variety of different names, and it is only relatively recently that rational taxonomic proposals for these organisms have emerged. Delineation of species within the genus is still the subject of ongoing research, and there is a particular need for a rapid and reliable method of assigning new isolates to individual species.

Notwithstanding the taxonomic confusion, it is clear that acinetobacters are common, free-living saprophytes found in soil, water and foods. They are also ubiquitous organisms in the clinical environment, where they can be isolated as commensals from the skin of hospital staff and patients, and they have increasingly been recognised as important nosocomial pathogens involved in outbreaks of hospital infection, particularly in high-dependency or intensive care units. The contributions that follow in this book are concerned primarily with the diagnosis, management and control of *Acinetobacter* infections. The aim of this chapter is to present a broad introductory overview of the biology of the genus *Acinetobacter* in its entirety and to highlight the most important features.

2.2 TAXONOMY

2.2.1 History of the Genus

Bacteria now classified as members of the genus *Acinetobacter* have suffered a long history of taxonomic change. At least 15 different 'generic' names have been used to describe these organisms, the most well-known of which are *Bacterium anitratum* (Schaub and Hauber, 1948), *Herellea vaginicola* and *Mima polymorpha* (De Bord, 1939), *Achromobacter, Alcaligenes, Micrococcus calcoaceticus* and 'B5W' (Juni, 1978), and *Moraxella glucidolytica* and *Moraxella lwoffii* (Piéchaud et al., 1956; Brisou, 1957).

The original concept of the genus *Acinetobacter*, as first proposed by microbiologists working in France (Brisou and Prévôt, 1954; Piéchaud et al., 1956; Brisou, 1957), included a heterogeneous collection of non-motile, Gram-negative, oxidase-positive (*Moraxella*) and oxidase-negative saprophytes that could be distinguished from other bacteria by their lack of pigmentation (Ingram and Shewan, 1960). The nutritional studies of Baumann et al. (1968) showed clearly that the oxidase-negative strains differed from the oxidase-positive strains of the genus *Moraxella*, and in 1971 the Subcommittee on the Taxonomy of *Moraxella* and Allied Bacteria recommended that the genus *Acinetobacter* should include only the oxidase-negative strains (Lessel, 1971). This division has been supported by the use of transformation tests (Juni, 1972), which now form the basis for inclusion of individual isolates within the genus.

Bergey's Manual of Systematic Bacteriology classified the genus *Acinetobacter* in the family Neisseriaceae (Juni, 1984), with one species, *A. calcoaceticus*, comprising two varieties, var. *anitratus* (formerly *Herellea vaginicola*) and var. *lwoffii* (formerly *Mima polymorpha*). More recent taxonomic developments have resulted in the proposal that members of the genus should be classified in the new family Moraxellaceae, which includes *Moraxella*, *Acinetobacter*, *Psychrobacter* and related organisms (Rossau et al., 1991), and which constitutes a discrete phylometric branch in superfamily II of the Proteobacteria on the basis of 16S rRNA studies and rRNA:DNA hybridisation assays (van Landschoot et al., 1986; Rossau et al., 1989).

The genus *Acinetobacter* is now defined as including Gram-negative (but sometimes difficult to destain) coccobacilli, with a DNA G+C content of 39–47 mol%, that are strictly aerobic, non-motile, catalase-positive and oxidase-negative. Good growth occurs on complex media between 20° and 30°C without growth factor requirements, while nitrates are reduced only rarely. Crucially, extracted DNA is able to transform mutant strain BD413 *trpE27* to the wild-type phenotype. Most strains of *Acinetobacter* can grow in a simple mineral medium containing a single carbon and energy source such as acetate, lactate or pyruvate. Further details of the identification of members of the genus *Acinetobacter* can be found in the following chapter.

2.2.2 Current Taxonomic Status and Delineation of Species

Traditionally, a microbial species has been considered to be a group of strains that show a high degree of similarity in terms of their phenotypic properties. However, in recent years it has become accepted by microbial taxonomists that nucleic acid hybridisation and sequencing studies provide the best available and most rational methods for designating species and determining relationships between different organisms. Wayne et al. (1987) proposed a formal molecular definition of a species—i.e., that a species should include strains with approximately 70% or greater DNA-DNA relatedness and 5°C or less divergence values (Δ Tm). Genomic species which can be differentiated by phenotypic properties can then be given a formal species name.

Although early DNA hybridisation experiments with a nitrocellulose filter method had shown the genus *Acinetobacter* to be heterogeneous (Johnson et al., 1970), only two species, *A. calcoaceticus* and *A. lwoffii* were included in the *Approved Lists of Bacterial Names* (Skerman et al., 1980), and only one species was described in *Bergey's Manual of Systematic Bacteriology* (Juni, 1984). Extensive studies using the DNA relatedness criteria outlined above have resulted in the recognition, to date, of 19 DNA-DNA homology groups (genomic species) within the genus (Bouvet and Grimont, 1986; Nishimura et al., 1987; Bouvet and Jeanjean, 1989; Tjernberg and Ursing, 1989). There are some minor discrepancies in the numbering schemes proposed for genomic species by different laboratories. Seven of the genomic species have been given formal species names (Table 2.1). The species *A. radioresistens*, proposed by Nishimura et al. (1988), was shown to be equivalent to genomic species 12 of Bouvet and Grimont (1986) by Tjernberg and Ursing (1989). A definitive numbering scheme has yet to be finally agreed. Indeed, groups 1, 2 and 3 of Bouvet and Grimont (1986) and group 13 of Tjernberg and Ursing (1989) have been shown to have an extremely close relationship and are referred to by some research groups as the "*A. calcoaceticus*-*A. baumannii* complex" (Gerner-Smidt et al., 1991). Methods for identifying the various genomic species are described in detail in the following chapter.

2.3 ECOLOGY AND HABITATS

2.3.1 Species of Clinical Importance

In addition to carriage by patients and staff, various strains of *Acinetobacter* can be isolated from a wide range of both moist and dry inanimate sources in the hospital environment, including Formica™ table tops, dust, ventilatory equipment, humidifiers, mattresses, urinals and wash basins. The epidemiology and types of infections caused by *Acinetobacter* spp. are described in detail elsewhere in this volume, but numerous studies have now supported the original proposal that *A. baumannii* is the main genomic species associated with outbreaks of

TABLE 2.1

Comparison of the Delineation of *Acinetobacter* Genomic Species by Different Laboratories

	Genomic species number according to			
Species name	**References 1 and 2**	**Reference 3**	**References 4 and 5**	**Type strain**
A. calcoaceticus	1	1	1	ATCC 23055
A. baumannii	2	2	1	CIP 70.34
	3	3	nt	ATCC 19004
	ug	13	nt	ATCC 17903
A. haemolyticus	4	4	4	ATCC 17906
A. junii	5	5	nt	ATCC 17908
	6	6	4	ATCC 17979
A. johnsonii	7	7	3	ATCC 17909
A. lwoffii	8	8	2	ATCC 15309
	9	8	nt	ATCC 9957
	10	10	ug	ATCC 17924
	11	11	ug	ATCC 11171
A. radioresistens	(12)*	12	5	IAM 13186
	13	14	nt	ATCC 17905
	14	nt	nt	Bouvet 382
	15	nt	nt	Bouvet 240
	16	ug	nt	ATCC 17988
	17	nt	nt	Bouvet 942
	nt	15	nt	Tjernberg 151a

Note: ug, ungrouped; nt, not tested; *unpublished result.

References: (1) Bouvet and Grimont (1986); (2) Bouvet and Jeanjean (1989); (3) Tjernberg and Ursing (1989); (4) Nishimura et al. (1987); and (5) Nishimura et al. (1988).

nosocomial infection (Bouvet and Grimont, 1987). Thus, a typical study of 584 *Acinetobacter* isolates isolated from 420 patients at 12 different hospitals over a 12-month period (Seifert et al., 1993a) identified 426 (72.9%) strains as *A. baumannii,* with 208 *A. baumannii* isolates being recovered from respiratory tract specimens, 113 from blood cultures and central veinous lines, 70 from wound swabs and 35 from other miscellaneous specimens (Table 2.2). This large study also identified 158 isolates as belonging to species other than *A. baumannii,* of which the most common were *Acinetobacter* sp. 3 (55 isolates), *A. johnsonii* (29 isolates) and *A. lwoffii* (21 isolates).

Very little is known about the clinical significance of *Acinetobacter* spp. other than *A. baumannii,* and further detailed investigations are required. The ubiquitous occurrence of acinetobacters in the environment, and as commensals on human skin, means that such isolates in clinical specimens are often considered to be contaminants. Diagnosis of infection with 'unusual' *Acinetobacter* genomic species therefore often

TABLE 2.2

Distribution of *Acinetobacter* spp. Isolated from Clinical Sources (data taken from Seifert et al., 1993a)

Genomic species	Number of isolates from each specimen type: Tracheal aspirate	Blood culture /C.V. line	Wound swab	Other sites	Total
A. baumannii	208	113	70	35	426
sp. 3	6	24	10	15	55
A. johnsonii	0	19	2	8	29
A. lwoffii	2	12	2	5	21
A. junii	0	7	0	4	11
A. haemolyticus	0	1	2	6	9
sp. 10	0	6	1	2	9
sp. 11	1	0	1	2	4
sp. 12	0	3	0	0	3
sp. 6	0	1	0	0	1
Unidentified	1	5	2	8	16
Total	218	191	90	85	584

depends on clinical indications and repeated isolation of the same strain from a single patient. *Acinetobacter* spp. 3 and 13 (as defined by Tjernberg and Ursing, 1989) have been implicated in documented outbreaks of infection involving nosocomial spread (Dijkshoorn et al., 1993), while *A. johnsonii* has been associated with catheter-related bacteraemia (Seifert et al., 1993b). A study in Sweden found that *Acinetobacter* sp. 3 was predominant among clinical isolates (Tjernberg and Ursing, 1989). It is worth reemphasising the close relationship between groups 1 (*A. calcoaceticus*), 2 (*A. baumannii*) and 3 of Bouvet and Grimont (1986) and group 13 of Tjernberg and Ursing (1989). This "*A. calcoaceticus-A. baumannii* complex" (Gerner-Smidt et al., 1991; Kämpfer et al., 1993) contains isolates that are mostly glucose-acidifying, and the "complex" therefore corresponds quite well with the *A. calcoaceticus* var. *anitratus* designation that was often used before the latest taxonomic developments. The majority of glucose-negative, non-haemolytic strains found in clinical specimens seem to identify mainly as *A. lwoffii*, *A. johnsonii* or *Acinetobacter* sp. 12, and it seems that these species are natural inhabitants of human skin. Most of the haemolytic isolates identify as *A. haemolyticus* or *Acinetobacter* sp. 6. Other groups seem to be implicated only occasionally in human infections. Overall, however, knowledge of the clinical importance of genomic species other than *A. baumannii* is poor.

In conclusion, although *A. baumannii* appears to be the *Acinetobacter* species of greatest clinical importance, repeated isolation of another genomic species (particularly one belonging to the *A. calcoaceticus*-*A. baumannii* complex) from a patient should be a cause for suspicion of infection, especially if clinical symptoms are also present.

2.3.2 Species from the Environment

Most taxonomic studies within the genus *Acinetobacter* have been performed with clinical isolates. However, acinetobacters are ubiquitous organisms that can be obtained easily from soil, water and sewage with appropriate enrichment techniques (Baumann, 1968), and members of the genus are also found in a variety of foodstuffs, including eviscerated chicken carcasses, various poultry and other meats, and milk products (Koburger, 1964; Barnes and Thornley, 1966; Eribo and Jay, 1985). Although it is known that acinetobacters are responsible for the economically important spoilage of foods such as bacon, chickens, eggs and fish, even when stored under refrigerated conditions (Shewan et al., 1960; Thornley et al., 1960; Gardner, 1971; Jay, 1982) or following irradiation treatment (Firstenberg-Eden et al., 1980), there have been surprisingly few studies in which *Acinetobacter* isolates from environmental sources have been grouped according to the latest taxonomic criteria.

So far as food isolates are concerned, a study by Gennari and Lombardi (1993) compared 170 *Acinetobacter* isolates from various food sources (including fresh and spoiled meat and fish, vegetables, raw milk and cheese) with 83 isolates from clinical sources. The results (Table 2.3) demonstrated a clear difference in the distribution of genomic species between the food and clinical isolates, with *A. johnsonii* (sp. 7) and *A. lwoffii* (spp. 8/9) predominating in foods. A similar study of non-motile, non-fermentative Gram-negative bacteria from foods (Shaw and Latty, 1988) also found a cluster of acinetobacters resembling *A. johnsonii*, and the ubiquitous occurrence of this genomic species in the environment has also been demonstrated in several studies of polyphosphate-accumulating acinetobacters found in wastewater treatment plants (Kämpfer et al., 1992; Knight et al., 1993). This contrasts with the predominance of *A. baumannii* and *Acinetobacter* sp. 3 in the vast majority of studies involving clinical isolates.

In conclusion, it appears that there is a significant population difference between the acinetobacters found in clinical and other environments. Each population appears to be characterised by predominant groups of genomic species—spp. 2 (*A. baumannii*) and 3 in clinical environments, and spp. 7 (*A. johnsonii*) and 8/9 (*A. lwoffii*) in other environments. Other genomic species appear to constitute only minority components of the different populations. Some interchange between populations is clearly possible for those groups that are capable of

TABLE 2.3

Comparison of the Distribution of *Acinetobacter* Genomic Species Found in Isolates From Food and Clinical Sources (data taken from Gennari and Lombardi, 1993)

Genomic species identified	Percentage of isolates Food sources (n = 170)	Clinical sources (n = 83)
1 *A. calcoaceticus*	3.5	0
2/13* *A. baumannii*	5.3	36.1
3	3.5	13.3
4 *A. haemolyticus*	0.5	4.8
5 *A. junii*	2.9	7.2
6	0	1.2
7 *A. johnsonii*	45.9	6.0
8/9/15* *A. lwoffii*	14.7	7.2
10	1.8	6.0
11	8.8	1.2
12 *A. radioresistens*	0.5	3.6
13/14*	0	8.4
16	2.4	0
Ungrouped	10.0	4.8

*According to Tjernberg and Ursing (1989).

growing at both 37°C and lower environmental temperatures, and this fact may have implications for the spread of antibiotic resistance genes (see Section 3.4).

From a clinical epidemiology viewpoint, it is apparent that the isolation of strains belonging to spp. 2 or 3 in environmental cross-infection studies should be considered to be of greater epidemiological significance than the detection of genomic species associated mostly with non-human sources.

2.4 GENETICS AND MOLECULAR BIOLOGY

2.4.1 Mechanisms of Chromosomal Gene Transfer

All three of the major modes of chromosomal gene transfer have been demonstrated in *Acinetobacter* spp.

Transformation

This process was demonstrated for the first time by Juni and Janik (1969) with *A. calcoaceticus* strain BD4 and its microencapsulated *trpE27* mutant derivative BD413. Transformation of BD413 *trpE27* to the wild-type phenotype now forms the basis of the genetic test for identifying members of the genus (Juni, 1972). Strain BD413 has an exceptionally high efficiency of transformation (0.5–0.7%), and it is necessary merely to spread BD413 cells, freshly grown in broth, together with suitable donor DNA on selective agar plates and then incubate in order to obtain transformants. The transformation process in this strain is so efficient that it facilitates a whole range of advanced molecular procedures that allow detailed fine-structure genetic investigations of the chromosomal organisation and metabolic regulation of this strain (Ornston and Neidle, 1991).

However, in general, natural genetic competence to act as a recipient strain for transformation seems to be a somewhat rare phenomenon amongst *Acinetobacter* spp. (Juni, 1978). In contrast to strain BD413, where competence occurs throughout the growth cycle of the cell with a peak early in the exponential growth phase, some other strains (e.g., strain EBF65/65) do not seem to become naturally competent at any stage in their growth cycle, while others (e.g., strain NCIB 8250) show limited competence only as a sharp peak at an early stage of batch culture (Vivian, 1991).

Transduction

A large number of bacteriophages active against specific strains of *Acinetobacter* have been isolated (Ackermann et al., 1994). Typically, most *Acinetobacter* phages are lytic and are restricted in their host range to the original strain of *Acinetobacter* on which they were isolated. One temperate phage (designated P78) which lysogenises its host and is capable of mediating low-frequency generalised transduction has been described (Herman and Juni, 1974). However, phage P78 is specific for its host strain and failed to lysogenise 389 other independently isolated strains of *Acinetobacter*, including strain BD413. One possible explanation for the narrow host specificity observed with *Acinetobacter* phages may be the large number of different surface antigens found in this genus (Marcus et al., 1969). As a partial consequence, neither P78 nor any other bacteriophage has been used extensively for genetic studies in *Acinetobacter* spp.

Conjugation

The first report of conjugation in *Acinetobacter* spp. involved use of the broad host-range plasmid RP4 to mobilise the chromosome of strain EBF65/65 (Towner and Vivian, 1976), and this was followed by a report of conjugation involving the naturally occurring *Acinetobacter* plasmid pAV1 (Hinchliffe and Vivian, 1980). Chromosomal transfer occurs in both of these systems only on solid surfaces and not in liquid matings. Very little is known about the precise nature and events involved in mobilisation of the chromosome mediated by RP4 and pAV1, but these systems have been used successfully to map a number of different mutations on a circular chromosomal linkage group in strain EBF65/65 (Towner, 1978; Vivian, 1991).

2.4.2 Genetic Organisation and Regulation

Studies aimed at defining the gross topological structures of bacterial genetic material are currently somewhat unfashionable. So far as the genus *Acinetobacter* is concerned, there have been few significant advances since Towner (1978) demonstrated that strain EBF65/65 had a circular chromosomal linkage map. A total of 29 genetic loci have been mapped on the chromosome of strain EBF65/65 (Vivian, 1991), but there have been insufficient studies to discern any consistent gross overall features in the arrangement of genes on the chromosome of this strain.

In complete contrast, although data on the gross chromosomal structure are lacking, knowledge of the fine structure of certain specific regions of the chromosome is quite well advanced for the highly competent *Acinetobacter* strain BD4. Much of this information has been obtained by means of transformation-mediated genetic exchange, with particular attention being focused on the organisation and regulation of genes concerned with the tryptophan biosynthetic and the ß-ketoadipate degradative pathways (Haspel et al., 1991; Ornston and Neidle, 1991). Striking similarities exist between the individual genes from *Acinetobacter* spp. and *Pseudomonas* spp. in these pathways, but considerable gene rearrangements seem to have occurred during the evolutionary process. Although the complete regulatory circuits for the tryptophan biosynthetic pathway have yet to be fully resolved in *Acinetobacter* spp. (Haspel et al., 1991), detailed studies of the ß-ketoadipate pathway have indicated that *Acinetobacter* strain BD4 exhibits induction patterns unlike those found in *Pseudomonas* spp. (Ornston and Neidle, 1991). However, an important characteristic shared with *Pseudomonas* spp. is that genes for physiologically interdependent steps in the ß-ketoadipate degradative pathway tend to be linked together in supraoperonic clusters, a feature that may have important regulatory implications (Ornston and Neidle, 1991).

2.4.3 Plasmids and Transposons

Plasmids and transposons play an important role in the biology of most prokaryotic organisms, and *Acinetobacter* spp. appear to be no exception to this generalisation. Crucial to this role is the location on plasmids and transposons of genes encoding a large and useful range of phenotypic properties, including antibiotic resistance and unusual metabolic activities. Several studies have reported that >80% of *Acinetobacter* isolates carry multiple indigenous plasmids of variable molecular size (e.g., Gerner-Smidt, 1989; Seifert et al., 1994), although other workers report problems in isolating plasmid DNA from *Acinetobacter* spp., often because of unappreciated difficulties in lysing the cell wall of these organisms.

Most indigenous plasmids from *Acinetobacter* spp. seem to be relatively small (<23 kb), and therefore probably lack conjugative functions. Apart from antibiotic resistance (see below), genes encoding resistance to heavy metals (e.g., Kholodii et al., 1993) and important metabolic steps in the degradation of environmental pollutants, such as polychlorinated biphenyls (PCBs), have been shown to be carried on plasmids and transposons in *Acinetobacter* spp. (reviewed by Towner, 1991). A range of cloning and shuttle vectors for *in vitro* genetic manipulation experiments in *Acinetobacter* spp. have also been described (Ditta et al., 1985; Singer et al., 1986; Hunger et al., 1990; Gutnick et al., 1991).

From the clinical viewpoint of this book, most interest inevitably focuses on plasmids and transposons carrying genes encoding resistance to antibiotics. Although many clinical isolates of *A. baumannii* show widespread and increasing resistance to a whole range of antibiotics (see Chapters 7 and 8), there have been very few studies in which plasmid-mediated transfer of resistance genes has been demonstrated. As indicated above, this may partly reflect a lack of conjugative functions on indigenous plasmids, but in addition, as postulated by Towner (1991), failure to observe transfer of resistance may simply reflect the absence of a suitable test system for detecting such transfer. For historical reasons, attempts to transfer plasmids from clinical isolates of any Gram-negative species have tended to use *Escherichia coli* K12 as a recipient strain. Complex and varied transfer frequencies of standard plasmids belonging to different incompatibility groups have been observed between *Acinetobacter* strain EBF65/65 and *E. coli* K12, and a number of these plasmids required an additional mobilising plasmid for retransfer to occur (Chopade et al., 1985). Accordingly, it is not surprising that most reported cases of indigenous transmissible antibiotic resistance from *Acinetobacter* spp. have been associated with plasmids belonging to broad host-range incompatibility groups (Towner, 1991). Nevertheless, there remains a need to rigorously assess the potential for mobilisation of antibiotic resistance genes to and from different genomic species of *Acinetobacter*.

This latter aspect is of particular interest, since studies to date indicate clearly that while there is a pool of plasmid-mediated genetic information that is confined largely to *Acinetobacter* spp., and a similar (but separate) set of information in clinical isolates of enteric bacteria, there is an 'intermediate' group of plasmids that can cross the boundaries between these and other distinct genetic pools. The observation that much of the antibiotic resistance found in clinical isolates of *Acinetobacter* spp. seems to result from the presence of enzymic resistance mechanisms that are essentially the same as those found in other genera of bacteria is evidence that this interchange can occur. Similarly, as discussed in Section 2.3, while some *Acinetobacter* genomic species seem to be confined largely to the human clinical environment, and others to soil, water or foods, there also exists an intermediate set of genomic species (e.g., spp. 7, 8 and 9) that are capable of adapting to both types of environment. When the high transformation frequencies observed with certain strains are taken into account, it seems possible that *Acinetobacter* spp. may offer a conduit by which novel antibiotic resistance genes from antibiotic-producing soil microorganisms might find their way into important human pathogenic bacteria in the clinical environment. Transposons probably play an important role, in conjunction with integrons (Collis et al., 1993), in ensuring that particular novel genes can become established in a new gene pool, even if the plasmid vectors that transferred them are unstable, and there have been several reports of chromosomally located transposons carrying multiple antibiotic resistance genes in clinical isolates of *Acinetobacter* spp. (Towner, 1991).

2.5 OVERVIEW OF BIOCHEMISTRY AND PHYSIOLOGY

The main biochemical characteristics that are used for identification of the genus *Acinetobacter* have been mentioned briefly in Section 2.2.1 and are discussed in more detail in the following chapter. This section summarises the major metabolic features of the genus and highlights the more interesting or unique physiological aspects. Early work on the metabolism of the genus was reviewed comprehensively by Juni (1978). The subsequent steady flow of papers describing enzymes, cofactors, metabolic pathways, products and other physiological aspects has, in general, confirmed that *Acinetobacter* spp. are broadly typical of other Gram-negative eubacteria, but that they have a number of distinctive metabolic features which support the versatile lifestyle of this genus. It is important to emphasise that many *Acinetobacter* strains used in biochemical and physiological studies are only now starting to become identified to the genomic species level, and that many of the detailed observations described in this section have been made on the basis of observations with only very limited numbers of strains.

2.5.1 Metabolism and Enzymology

Most strains of *Acinetobacter* do not have any growth factor requirements and resemble the saprophytic pseudomonads in being able to use any one of a large number of organic carbon and energy sources in an otherwise mineral medium. Although the utilisation of carbohydrates is relatively uncommon, a major feature of the genus is that many strains can metabolise a wide range of compounds, including aliphatic alcohols, some amino acids, decarboxylic and fatty acids, unbranched hydrocarbons, sugars, and many relatively recalcitrant aromatic compounds such as benzoate, mandelate, *n*-hexadecane, cyclohexanol and 2,3-butanediol. Acinetobacters have therefore been used widely for studying a variety of unusual biochemical pathways, and may also have a role to play in degrading a variety of pollutants and industrial products (see Section 2.6.4).

So far as is known, metabolism centres on a Krebs tricarboxylic acid cycle with a rather unusual pattern of regulation in which the activities of a number of the constituent enzymes are all modulated by AMP. This distinctive feature may act in a concerted manner to direct metabolic flux through the cycle, and such multipoint control of the cycle may be a significant feature in the regulation of energy metabolism in *Acinetobacter* spp. Numerous complex catabolic pathways for the degradation of aromatic and acyclic compounds (as well as alkanes, amino acids and related compounds) feed into the tricarboxylic acid cycle, and various biosynthetic pathways, such as those involved in gluconeogenesis and amino acid synthesis, are subtly influenced by the need to integrate with the overall metabolism of the organism (Towner et al., 1991).

As indicated previously, most strains are unable to utilise glucose as a carbon source, but many strains are able to acidify media containing sugars, including glucose, via an aldose dehydrogenase. This property was considered previously to be of major taxonomic significance, but has now been supplanted in importance by the developments described in Section 2.2. Many *Acinetobacter* strains accumulate poly-ß-hydroxybutyrate and wax esters as energy reserves, and some strains accumulate polyphosphate, either as an energy reserve or as a phosphate reserve. Although most acinetobacters are unable to reduce nitrate to nitrite in conventional assays, some strains can use both nitrate and nitrite as nitrogen sources by means of an assimilatory nitrate reductase. All acinetobacters are oxidase-negative, since they lack cytochrome *c*, but they do contain cytochromes *a* and *b*.

A wide range of extracellular enzymes are produced by acinetobacters, including lipases, esterases, amylase, collagenase, endo-ß-1,6-glucanase, and at least one ß-lactamase. There are several interesting periplasmic enzymes, including an insulin-cleaving metalloproteinase, and numerous interesting membrane-bound enzymes, including an NAD-independent malate dehydrogenase, various proteases, leucine aminopeptidase, alanine aminopeptidase, and NAD-independent aldehyde, alcohol, lactate and mandelate dehydrogenases (Towner et al., 1991). So far as cofactors are concerned, elucidation of the biological role of pyrroloquinoline quinone (PQQ), particularly in respect of its function in glucose dehydrogenase activity, has resulted largely from studies with various strains of *Acinetobacter* (Duine, 1991).

An important difference from many other organisms is that *Acinetobacter* spp. cannot incorporate extracellular thymine or thymidine into DNA because of a lack of the enzymes thymidine phosphorylase, nucleoside deoxyribosyltransferase and thymidine kinase. An additional special feature of at least one member of the genus is the ability to grow on malonate, and there has also been a report that some acinetobacters isolated from the surfaces of soybean nodules (where there may be appreciable concentrations of hydrogen because of the action of nitrogenase) can oxidise hydrogen. Further details of these and other metabolic curiosities can be found in Towner et al. (1991).

2.5.2 Energetics of Growth

Several strains of *Acinetobacter* have been examined systematically to determine the effects of variables such as substrate, temperature, oxygen status, growth rate and nutrient limitation on growth yields and composition (reviewed by Towner et al., 1991). Substrate inhibition has been found to be important in some cases, but growth yields and respiratory efficiencies have been shown to be rather low in batch and continuous culture experiments. This may be related to the fact that the effective P/O ratio is only about one under most conditions, although there may be two or more sites of oxidative phosphorylation (Fewson, 1985). Failure to make effective use of all the potential synthesis of ATP is consistent with the suggestion that the redox chain may be divided into two equivalent sites of energy conservation (Ensley et al., 1981), and results of experiments to determine H^+/O quotients also seem to suggest that there are at least two sites of oxidative phosphorylation (see Towner et al., 1991). There is some evidence that strains which cannot use glucose as a sole carbon and energy source can still dehydrogenate glucose to energise active transport systems and provide ATP for biosynthesis (Kitagawa et al., 1986a; b).

The nature of the electron transfer chain in acinetobacters is still not elucidated completely, despite work over several decades by a number of different groups (reviewed by Towner et al., 1991). A cytochrome *o*-containing oxidase and a cytochrome b_{554} predominate under conditions of high aeration, while a cytochrome *d*-containing oxidase predominates under oxygen-limited conditions. Cytochrome b_{562} is involved in linking glucose dehydrogenase to the main electron transfer chain, and several strains are known to contain cytochrome P450. No evidence has been found in acinetobacters for the existence of cytochrome *c*, and this is in agreement with the accepted oxidase-negative status of the genus.

2.5.3 Transport Systems

Acinetobacters are known to contain active transport systems for a number of different substrates, and in some strains these transport systems have been shown to be energised by glucose dehydrogenation (van Schie et al., 1985; Kitagawa et al., 1986a; b). Fewson (1991) indicated that some aromatic compounds are also taken up by specific transport systems, and at least two different phosphate transport systems have been identified in a strain of *A. johnsonii* capable of accumulating polyphosphate from wastewater (van Veen et al., 1993).

The ability of a bacterium to obtain the necessary iron for growth in the human body is now recognised to be an important virulence determinant, and some strains of *Acinetobacter* have been shown to produce siderophores and iron-repressible outer-membrane receptor proteins (Smith et al., 1990; Echenique et al., 1992; Actis et al., 1993). These observations may have important implications when attempting to explain the virulence of certain strains, and it is clear that the entire question of uptake of various compounds and molecules into acinetobacters (and related efflux systems) is an important area for further research effort if the physiology and pathogenicity of this genus is to be elucidated adequately.

2.6 CLINICAL, INDUSTRIAL AND COMMERCIAL APPLICATIONS

The general biochemical versatility of members of the genus has resulted in *Acinetobacter* spp. receiving considerable attention as possible industrial microorganisms. A number of patent applications incorporating these organisms in various industrial processes have been awarded. Some of the suggested uses for *Acinetobacter* spp. and their products are summarised in Table 2.4. The most important of these are described in more detail in the following sections.

TABLE 2.4

Major Suggested Uses for *Acinetobacter* spp. and Their Products

Treatment of wastewaters and effluents
- Phosphate removal
- Degradation of petrochemicals
- Breakdown of organic pollutants
Production of biopolymers and biosurfactants
- For prevention of dental plaque
- For use in paper-making and other industries
- For efficient emulsification of oil waste pollutants
- For incorporation in cosmetics, detergents and shampoos
Clinical uses
- Production of glutaminase-asparaginase
- Production of L(-)-carnitine
Biomass production
- Single cell protein production
Manganese leaching from ores
Production of immune adjuvants

2.6.1 Clinical Applications

Certain strains of *Acinetobacter* produce glutaminase-asparaginase, an enzyme that has been considered for possible use as an anti-cancer agent (e.g., Kien et al., 1985; Ammon et al., 1988) on the basis that cancerous cells may be unable to synthesise sufficient amounts of certain non-essential amino acids, and that such cells will be killed selectively if the circulatory level of such amino acids is reduced. Parenteral administration of L-asparaginase in experimental animals and human volunteers has been shown to be associated with regression of lymphomas and leukaemias. The *Acinetobacter* enzyme utilises both L-glutamine and L-asparagine as substrates, and a number of studies have been aimed at reducing immunological complications and increasing the circulatory life of the enzyme.

L(–)-carnitine is a registered drug for the treatment of carnitine deficiency syndromes, with uses in replacement therapy for haemodialysis patients, for therapy and prevention of heart diseases linked to disturbed fat metabolism, and for related indications such as hyperlipoproteinaemia. Unfortunately, inexpensive racemic carnitine can no longer be used for treatment since the D-enantiomer has been shown to have harmful side effects. Consequently, a variety of processes have been developed for the production of enantiomerically pure L(–)-carnitine. Several strains of *Acinetobacter* have been described which are characterised by their unique ability to preferentially degrade D(+)-carnitine in a racemic mixture under aerobic conditions, a process which can then be followed by recovery of the pure L(–)-carnitine (Kleber, 1991).

A third major clinical application, this time for an *Acinetobacter* bioproduct, concerns the improvement of oral hygiene by the removal of dental plaque. Emulsan is the name given to the extracellular form of a polyanionic, cell-associated, heteropolysaccharide produced by *Acinetobacter* strain RAG-1 (Gutnick, 1987), while dental plaque is a mucilaginous gelatinoid type of deposit which forms on the surface of teeth and which is invaded readily by bacteria, leading eventually to dental caries. U.S. Patent No. 4,619,825, filed on behalf of the Colgate-Palmolive Company, describes the use of emulsan-containing formulations for either mouthwash or toothpaste which have the effect of significantly reducing dental plaque formation and enhancing anti-caries activity.

2.6.2 Other Applications of Bioproducts

Apart from its use in the removal of dental plaque, emulsan has a whole range of potential applications in the petroleum industry, including viscosity reduction during pipeline transport following formation of heavy oil:water emulsions, and production of fuel oil:water emulsions for direct combustion. The affinity of purified emulsan for the oil:water interface also has implications for the stability of oil emulsions during transport and storage, and also for their biodegradability following accidental spillage (Gutnick et al., 1991).

A second important polysaccharide produced by a strain of *Acinetobacter* has been shown to bind to inorganic materials such as calcium carbonate (limestone). This polysaccharide has been termed biodispersan (Rosenberg et al., 1988) and is capable of dispersing limestone in water. Since limestone is used in a wide range of industrial processes, purified biodispersan is considered to have potential applications in manufacturing processes producing common products such as paper, paints and ceramics.

2.6.3 Biotransformations

The ability of acinetobacters to metabolise a wide range of aromatic, alicyclic and other compounds has resulted in a number of possible commercial biotransformation processes, including the production of L(–)-carnitine (see Section 2.6.1). Biotransformation processes can be achieved with either whole cells or purified enzymes, although a disadvantage implicit in the use of whole cells is the potentially rapid consecutive degradation of the products of interest. Many of the unusual biochemical pathways utilised by *Acinetobacter* spp. are still in the process of being elucidated, but it is already clear that it may be possible to accumulate intermediate products of commercial interest by constructing mutant strains that are blocked at the appropriate point of a particular biochemical pathway. In this respect, compared with other microorganisms of similar metabolic activity, acinetobacters have an important advantage in that they are easy to isolate, cultivate and manipulate genetically in the laboratory.

2.6.4 Pollution Control

Phosphate removal from wastewater is an important feature of sewage treatment plants in order to minimise the problem of eutrophication. Although the addition of various inorganic compounds (e.g., calcium, iron or aluminium salts) to activated sewage sludge, followed by settlement, can be used to remove phosphates, the ingredients for the process are expensive, and large quantities of chemical waste are generated. Examination of activated sludge samples from sewage treatment plants has shown that acinetobacters form a significant part of the constituent microflora. Although their precise role, and its importance, remains a matter of some debate, it can be shown in the laboratory that such strains can accumulate polyphosphates, thereby forming the potential basis of an inexpensive biological method for the direct removal of phosphate from wastewaters.

In addition to their potential role in the removal of phosphate, particular strains of *Acinetobacter* have the ability to degrade numerous hazardous and unpleasant waste and residue pollutant compounds produced as by-products of commercial processes. Examples of potential or actual applications of acinetobacters in the area of pollution control are listed in Table 2.5. In a number of cases, total degradation occurs by the synergic action of complex microbial communities in which acinetobacters form an important component. Further details can be found in the comprehensive review of Fewson (1991).

2.7 SUMMARY

This book is concerned largely with the diagnosis, severity and treatment difficulties associated with *Acinetobacter* infections, and

TABLE 2.5

Examples of Actual or Potential Target Pollutants Capable of Partial or Total Degradation by *Acinetobacter* spp.

Cresols	Phenols
Phthalic acid esters	Aniline
Diphenylethylene	Hydroxybenzoate
Anisate	Benzene
Petroleum products	Salicylate
Crude oil	Phenylacetate
Creosote components	Lignins
Linear alkylbenzene sulphonates	Hydroxyquinolone
Asphalt components	2,4-Dichlorophenoxyacetate (2-4-D)
4-Chlorobiphenyl	4-Chlorobenzoate
Polychlorinated biphenyls (PCBs)	Methoxychlor
Aroclor 1242	Mecoprop

particularly those difficulties related to the high rates of resistance now found in this genus to the major groups of antibiotics. Although the significant pathogenic role of *Acinetobacter* is confined to nosocomial infections, it is important to note that these organisms are ubiquitous in both the clinical and non-clinical environments. Apart from the possible role of acinetobacters as a reservoir of antibiotic resistance genes, their metabolic versatility means that they may play an important role in a variety of commercially important industrial processes, as well as in the biodegradation of a wide range of environmental pollutants. While much remains to be learned about the lifestyles of the different members of the genus, it is clear that some facets of their behaviour are unique, while others provide new insights into prokaryotic behaviour in general. Members of the genus *Acinetobacter* have suffered a long history of extensive taxonomic change which has inhibited a proper appreciation and understanding of their significant biological properties and pathogenic importance. Although the delineation of species within the genus is still the subject of ongoing discussion and research, it appears that a satisfactory generic classification of *Acinetobacter* is at last at the disposal of microbiologists, and this should form a rational scientific foundation for the numerous important investigative studies required in the future.

REFERENCES

Ackermann, H.W., Brochu, G. and Konjin, H.P.E. (1994). Classification of *Acinetobacter* phages. *Arch. Virol.* 135, 345–354.

Actis, L.A., Tolmasky, M.E., Crosa, L.M. and Crosa, J.H. (1993). Effect of iron-limiting conditions on growth of clinical isolates of *Acinetobacter baumannii*. *J. Clin. Microbiol.* 31, 2812–2815.

Ammon, H.L., Weber, I.T., Wlodawer, A., Harrison, R.W., Gilliland, G.L., Murphy, K.C., Sjolin, L. and Roberts, J. (1988). Preliminary crystal studies of *Acinetobacter glutaminasificans* glutaminase-asparaginase. *J. Biol. Chem.* 263, 150–156.

Barnes, E.M. and Thornley, M.J. (1966). The spoilage of eviscerated chickens stored at different temperatures. *J. Food Technol.* 1, 113–119.

Baumann, P. (1968). Isolation of *Acinetobacter* from soil and water. *J. Bacteriol.* 96, 39–42.

Baumann, P., Doudoroff, M. and Stanier, R.Y. (1968). A study of the *Moraxella* group. II. Oxidase negative species (genus *Acinetobacter*). *J. Bacteriol.* 95, 1520–1541.

Bouvet, P.J.M. and Grimont, P.A.D. (1986). Taxonomy of the genus *Acinetobacter* with the recognition of *Acinetobacter baumannii* sp.nov., *Acinetobacter haemolyticus* sp.nov., *Acinetobacter johnsonii* sp.nov. and *Acinetobacter junii* sp.nov. and emended descriptions of *Acinetobacter calcoaceticus* and *Acinetobacter lwoffii*. *Int. J. System. Bacteriol.* 36, 228–240.

Bouvet, P.J.M. and Grimont, P.A.D. (1987). Identification and biotyping of clinical isolates of *Acinetobacter*. *Ann. Inst. Pasteur Microbiol.* 138, 569–578.

Bouvet, P.J.M. and Jeanjean, S. (1989). Delineation of new proteolytic genospecies in the genus *Acinetobacter*. *Res. Microbiol.* 140, 291–299.

Brisou, J. (1957). Contribution à l'étude des Pseudomonadaceae. Précisions taxonomiques sur le genre *Acinetobacter*. *Ann. Inst. Pasteur* 93, 397–404.

Brisou, J. and Prévôt, A.R. (1954). Études de systématique bacterienne. X. Révision des spéces réunies dans le genre Achromobacter. *Ann. Inst. Pasteur* 86, 722–728.

Chopade, B.A., Wise, P.J. and Towner, K.J. (1985). Plasmid transfer and behaviour in *Acinetobacter calcoaceticus* EBF65/65. *J. Gen. Microbiol.* 131, 2805–2811.

Collis, C.M., Grammaticopoulos, G., Briton, J., Stokes, H.W. and Hall, R.M. (1993). Site-specific insertion of gene cassettes into integrons. *Mol. Microbiol.* 9, 41–52.

De Bord, G.G. (1939). Organisms invalidating the diagnosis of gonorrhea by the smear method. *J. Bacteriol.* 38, 119.

Dijkshoorn, L., Aucken, H.M., Gerner-Smidt, P., Kaufmann, M.E., Ursing, J. and Pitt, T.L. (1993). Correlation of typing methods for *Acinetobacter* isolates from hospital outbreaks. *J. Clin. Microbiol.* 31, 702–705.

Ditta, G., Schmidhauser, T., Yakobsen, E., Lu, P., Wu-liang, X., Findlay, D.R., Guiney, D. and Helinski, D. (1985). Plasmids related to the broad host range vector pRK290 useful for gene cloning and for monitoring gene expression. *Plasmid* 13, 149–153.

Duine, J.A. (1991). Energy generation and the glucose dehydrogenase pathway in *Acinetobacter*. In *The Biology of* Acinetobacter, pp. 295–312. K.J. Towner, E. Bergogne-Bérézin and C.A. Fewson (Eds). Plenum Press, New York.

Echenique, J.R., Arienti, H., Tolmasky, M.E., Read, R.R., Staneloni, R.J., Crosa, J.H. and Actis, L.A. (1992). Characterization of a high-affinity iron transport system in *Acinetobacter baumannii*. *J. Bacteriol.* 174, 7670–7679.

Ensley, B.D., Irwin, R.M., Carreira, L.A., Hoffmann, P.S., Morgan, P.S., Morgan, T.V. and Finnerty, W.R. (1981). Effects of growth substrates and respiratory chain composition on bioenergetics in *Acinetobacter* strain HO1-N. *J. Bacteriol.* 148, 508–513.

Eribo, B.E. and Jay, J.M. (1985). Incidence of *Acinetobacter* species and other Gram-negative, oxidase-negative bacteria in fresh and spoiled ground beef. *Appl. Environ. Microbiol.* 49, 256–257.

Fewson, C.A. (1985). Growth yields and respiratory efficiency of *Acinetobacter calcoaceticus*. *J. Gen. Microbiol.* 131, 865–872.

Fewson, C.A. (1991). Metabolism of aromatic compounds by *Acinetobacter*. In *The Biology of* Acinetobacter, pp. 351–390. K.J. Towner, E. Bergogne-Bérézin and C.A. Fewson (Eds). Plenum Press, New York.

Firstenberg-Eden, R., Rowley, D.B. and Shattuck, G.E. (1980). Factors affecting inactivation of *Moraxella-Acinetobacter* cells in an irradiation process. *Appl. Environ. Microbiol.* 40, 480–485.

Gardner, G.A. (1971). Microbiological and chemical changes in lean Wiltshire bacon during aerobic storage. *J. Appl. Bacteriol.* 34, 645–654.

Gennari, M. and Lombardi, P. (1993). Comparative characterization of *Acinetobacter* strains isolated from different foods and clinical sources. *Zentralbl. Bakteriol.* 279, 544–552.

Gerner-Smidt, P. (1989). Frequency of plasmids in strains of *Acinetobacter calcoaceticus*. *J. Hosp. Infect.* 14, 23–28.

Gerner-Smidt, P., Tjernberg, I. and Ursing, J. (1991). Reliability of phenotypic tests for identification of *Acinetobacter* species. *J. Clin. Microbiol.* 29, 277–282.

Gutnick, D.L. (1987). The emulsan polymer: perspectives on a microbial capsule as an industrial product. *Biopolymers* 23S, 223.

Gutnick, D.L., Allon, R., Levy, C., Petter, R. and Minas, W. (1991). Applications of *Acinetobacter* as an industrial microorganism. In *The Biology of* Acinetobacter, pp. 411–441. K.J. Towner, E. Bergogne-Bérézin and C.A. Fewson (Eds). Plenum Press, New York.

Haspel, G., Kishan, V. and Hillen, W. (1991). Organisation, potential regulatory elements and evolution of *trp* genes in *Acinetobacter*. In *The Biology of* Acinetobacter, pp. 239–249. K.J. Towner, E. Bergogne-Bérézin and C.A. Fewson (Eds). Plenum Press, New York.

Herman, N.J. and Juni, E. (1974). Isolation and characterisation of a generalised transducing bacteriophage for *Acinetobacter*. *J. Virol.* 13, 46–52.

Hinchliffe, E. and Vivian, A. (1980). Naturally occurring plasmids in *Acinetobacter calcoaceticus*: a P class R factor of restricted host range. *J. Gen. Microbiol.* 116, 75–80.

Hunger, M., Schmucker, R., Kishan, V. and Hillen, W. (1990). Analysis and nucleotide sequence of an origin of DNA replication in *Acinetobacter calcoaceticus* and its use for *Escherichia coli* shuttle plasmids. *Gene* 87, 45–51.

Ingram, M. and Shewan, J.W. (1960). Introductory reflections on the *Pseudomonas-Achromobacter* group. *J. Appl. Bacteriol.* 23, 373–378.

Jay, J.M. (1982). *Modern Food Microbiology*, 3rd ed. Van Nostrand, New York.

Johnson, J.L., Anderson, R.S. and Ordal, E.J. (1970). Nucleic acid homologies among oxidase-negative *Moraxella* species. *J. Bacteriol.* 101, 568–573.

Juni, E. (1972). Interspecies transformation of *Acinetobacter*: genetic evidence for a ubiquitous genus. *J. Bacteriol.* 112, 917–931.

Juni, E. (1978). Genetics and physiology of *Acinetobacter*. *Annu. Rev. Microbiol.* 32, 349–371.

Juni, E. (1984). Genus III. *Acinetobacter* Brisou et Prévôt 1954. In *Bergey's Manual of Systematic Bacteriology*, Vol. 1, pp. 303–307. N.R. Krieg and J.G. Holt (Eds). Williams & Wilkins, Baltimore.

Juni, E. and Janik, A. (1969). Transformation of *Acinetobacter calcoaceticus (Bacterium anitratum)*. *J. Bacteriol.* 98, 281–288.

Kämpfer, P., Bark, K., Busse, H.-J., Auling, G. and Dott, W. (1992). Numerical and chemotaxonomy of polyphosphate accumulating *Acinetobacter* strains with high polyphosphate:AMP phosphotransferase (PPAT) activity. *Syst. Appl. Microbiol.* 15, 409–419.

Kämpfer, P., Tjernberg, I. and Ursing, J. (1993). Numerical classification and identification of *Acinetobacter* genomic species. *J. Appl. Bacteriol.* 75, 259–268.

Kholodii, G.Y., Gorlenko, Z.M., Lomovskaya, O.L., Mindlin, S.Z., Yurieva, O.V. and Nikiforov, V.G. (1993). Molecular characterization of an aberrant mercury resistance transposable element from an environmental *Acinetobacter* strain. *Plasmid* 30, 303–308.

Kien, C.L., Anderson, A.J. and Holcenberg, J.S. (1985). Tissue nitrogen-sparing effect of high protein diet in mice with or without ascites tumour treated with *Acinetobacter* glutaminase-asparaginase. *Cancer Res.* 45, 4876–4882.

Kitagawa, K., Tateishi, A., Nakano, F., Matumoto, I., Morohoshi, T., Tanino, T. and Osui, T. (1986a). Generation of energy coupled with membrane-bound glucose dehydrogenase in *Acinetobacter calcoaceticus*. *Agric. Biol. Chem.* 50, 1453–1457.

Kitagawa, K., Tateishi, A., Nakano, F., Matumoto, I., Morohoshi, T., Tanino, T. and Osui, T. (1986b). Sources of energy and energy coupling reactions of the active transport systems in *Acinetobacter calcoaceticus*. *Agric. Biol. Chem.* 50, 2939.

Kleber, H.-P. (1991). Metabolism of trimethylammonium compounds by *Acinetobacter*. In *The Biology of* Acinetobacter, pp. 403–410. K.J. Towner, E. Bergogne-Bérézin and C.A. Fewson (Eds). Plenum Press, New York.

Knight, G.C., McDonnell, S.A., Seviour, R.J. and Soddell, J.A. (1993). Identification of *Acinetobacter* isolates using the Biolog identification system. *Lett. Appl. Microbiol.* 16, 261–264.

Koburger, S.A. (1964). Isolation of *Mima polymorpha* from dairy products. *J. Dairy Sci.* 47, 646.

Lessel, E.F. (1971). Minutes of the Subcommittee on the Taxonomy of *Moraxella* and Allied Bacteria. *Int. J. Syst. Bacteriol.* 21, 213–214.

Marcus, B.B., Samuels, S.B., Pittman, B. and Cherry, W.B. (1969). A serologic study of *Herellea vaginicola* and its identification by immunofluorescent staining. *Am. J. Clin. Pathol.* 52, 309–319.

Nishimura, Y., Ino, T. and Iizuka, H. (1988). *Acinetobacter radioresistens* sp.nov. isolated from cotton and soil. *Int. J. Syst. Bacteriol.* 38, 209–211.

Nishimura, Y., Kano, M., Ino, T., Iizuka, H., Kosako, Y. and Kaneko, T. (1987). Deoxyribonucleic acid relationship among the radioresistant *Acinetobacter* and other *Acinetobacter*. *J. Gen. Appl. Microbiol.* 33, 371–376.

Ornston, L.N. and Neidle, E.L. (1991). Evolution of genes for the ß-ketoadipate pathway in *Acinetobacter calcoaceticus*. In *The Biology of* Acinetobacter, pp. 201–237. K.J. Towner, E. Bergogne-Bérézin and C.A. Fewson (Eds). Plenum Press, New York.

Piéchaud, D., Piéchaud, M. and Second, L. (1956). Variétés protéolytiques de *Moraxella lwoffii* et de *Moraxella glucidolytica*. *Ann. Inst. Pasteur* 90, 517–522.

Rosenberg, E., Rubinovitz, C., Gottlieb, A., Rosenhak, S. and Ron, E.Z. (1988). Production of biodispersan by *Acinetobacter calcoaceticus* A2. *Appl. Environ. Microbiol.* 54, 317–322.

Rossau, R., van den Bussche, G., Thielemans, S., Segers, P., Grosch, H., Göthe, E., Mannheim, W. and De Ley, J. (1989). Ribosomal ribonucleic acid cistron similarities and deoxyribonucleic acid homologies of *Neisseria, Kingella, Eikenella, Alysiella,* and Centers for Disease Control Groups EF-4 and M-5 in the emended family Neisseriaceae. *Int. J. Syst. Bacteriol.* 39, 185–198.

Rossau, R., van Landschoot, A., Gillis, M. and De Ley, J. (1991). Taxonomy of Moraxellaceae fam.nov., a new bacterial family to accommodate the genera *Moraxella, Acinetobacter,* and *Psychrobacter* and related organisms. *Int. J. Syst. Bacteriol.* 41, 310–319.

Schaub, I.G. and Hauber, F.D. (1948). A biochemical and serological study of a group of identical unidentifiable Gram-negative bacilli in human sources. *J. Bacteriol.* 56, 379–385.
Seifert, H., Baginsky, R., Schulze, A. and Pulverer, G. (1993a). The distribution of *Acinetobacter* species in clinical culture materials. *Zentralbl. Bakteriol.* 279, 544–552.
Seifert, H., Strate, A., Schulze, A. and Pulverer, G. (1993b). Catheter-related bacteremia due to *Acinetobacter johnsonii*. *Clin. Infect. Dis.* 17, 632–636.
Seifert, H., Schulze, A., Baginski, R. and Pulverer, G. (1994). Plasmid DNA fingerprinting of *Acinetobacter* species other than *Acinetobacter baumannii*. *J. Clin. Microbiol.* 32, 82–86.
Shaw, B.G. and Latty, J.B. (1988). A numerical taxonomic study of non-motile non-fermentative Gram-negative bacteria from foods. *J. Appl. Bacteriol.* 65, 7–21.
Shewan, J.M., Hobbs, G. and Hodgkiss, W. (1960). The *Pseudomonas* and *Achromobacter* groups of bacteria in the spoilage of marine white fish. *J. Appl. Bacteriol.* 23, 463–468.
Singer, J.T., van Tuijl, J.J. and Finnerty, W.R. (1986). Transformation and mobilization of cloning vectors in *Acinetobacter* spp. *J. Bacteriol.* 165, 301–303.
Skerman, V.B.D., McGowan, V. and Sneath, P.A. (1980). Approved lists of bacterial names. *Int. J. Syst. Bacteriol.* 30, 225–420.
Smith, A.W., Freeman, S., Minett, W.G. and Lambert, P.A. (1990). Characterization of a siderophore from *Acinetobacter calcoaceticus*. *FEMS Microbiol. Lett.* 70, 29–32.
Thornley, M.J., Ingram, M. and Barnes, E.M. (1960). The effects of antibiotics and irradiation on the *Pseudomonas-Achromobacter* flora of chilled poultry. *J. Appl. Bacteriol.* 23, 487–498.
Tjernberg, I. and Ursing, J. (1989). Clinical strains of *Acinetobacter* classified by DNA-DNA hybridization. *Acta Pathol. Microbiol. Immunol. Scand.* 97, 596–605.
Towner, K.J. (1978). Chromosome mapping in *Acinetobacter calcoaceticus*. *J. Gen. Microbiol.* 104, 175–180.
Towner, K.J. (1991). Plasmid and transposon behaviour in *Acinetobacter*. In *The Biology of* Acinetobacter, pp. 149–167. K.J. Towner, E. Bergogne-Bérézin and C.A. Fewson (Eds). Plenum Press, New York.
Towner, K.J. and Vivian, A. (1976). RP4-mediated conjugation in *Acinetobacter calcoaceticus*. *J. Gen. Microbiol.* 93, 355–360.
Towner, K.J., Bergogne-Bérézin, E. and Fewson, C.A. (1991). *The Biology of* Acinetobacter. Plenum Press, New York.
van Landschoot, A., Rossau, R. and De Ley, J. (1986). Intra- and intergeneric similarities of the ribosomal ribonucleic acid cistrons of *Acinetobacter*. *Int. J. Syst. Bacteriol.* 36, 150–160.
van Schie, B.J., Hellingwerf, K.J., van Dijken, J.P., Elferinck, M.G.L., van Dijl, J.M., Kuenen, J.G. and Konings, W.N. (1985). Energy transduction by electron transfer via a pyrrolo-quinoline quinone dependent glucose dehydrogenase in *Escherichia coli, Pseudomonas aeruginosa* and *Acinetobacter calcoaceticus* (var. *lwoffi*). *J. Bacteriol.* 163, 493–499.
van Veen, H.W., Abee, T., Kortstee, G.J.J., Konings, W.N. and Zehnder, A.J.B. (1993). Characterization of two phosphate transport systems in *Acinetobacter johnsonii* 210A. *J. Bacteriol.* 175, 200–206.
Vivian, A. (1991). Genetic organisation of *Acinetobacter*. In *The Biology of* Acinetobacter, pp. 191–200. K.J. Towner, E. Bergogne-Bérézin and C.A. Fewson (Eds). Plenum Press, New York.

Wayne, L.G., Brenner, D.J., Colwell, R.R., Grimont, P.A.D., Kandler, O., Krichevsky, M.I., Moore, L.H., Moore, W.E.C., Murray, R.G.E., Stackebrandt, E., Starr, M.P. and Trüper, H.G. (1987). Report of the ad hoc committee on reconciliation of approaches to bacterial systematics. *Int. J. Syst. Bacteriol.* 37, 463–464.

Chapter 3

Acinetobacter — Microbiology

Lenie Dijkshoorn

CONTENTS

0-8493-9223-3/95/$0.00+$.50

3.1 INTRODUCTION

Over the decades, the taxonomy of the genus *Acinetobacter* has undergone many changes. The oldest still viable strain was described by Beijerinck (1911) as *Micrococcus calco-aceticus*. Thereafter, microorganisms now considered to belong to this genus were discovered independently by several microbiologists who gave them different names, such as *Bacterium anitratum* and *Moraxella lwoffii* (see Chapter 2 and Appendix I). In the absence of sufficient criteria for speciation, only one genus and one species were recognised in the Eighth Edition of *Bergey's Manual of Determinative Bacteriology* (Lautrop, 1974). Since 1986, a total of 19 genomic species (genospecies) have been established on the basis of DNA-DNA hybridisation studies (Bouvet and Grimont, 1986; Nishimura et al., 1988; Bouvet and Jeanjean, 1989; Tjernberg and Ursing, 1989). More DNA groups undoubtedly exist, as judged from DNA hybridisation studies of clinical isolates (Bouvet and Grimont, 1986; Bouvet and Jeanjean, 1989; Tjernberg and Ursing, 1989; Horrevorts et al., 1995). Seven genomic species (DNA groups) have names, while others are designated by numbers (see Chapter 2). The independently described groups 13–17 of Bouvet and Jeanjean (1989) and 13–15 of Tjernberg and Ursing (1989) have, in part, common designations, although only one group (no. 14 *sensu* Tjernberg and Ursing and no. 13 *sensu* Bouvet and Jeanjean) corresponds. To avoid confusion, groups are given the suffixes BJ and TU throughout this chapter.

DNA group 2 (*A. baumannii*) and the unnamed groups 3 and 13TU seem to be the species found most commonly in specimens from hospitalised patients. Groups 5 (*A. junii*), 7 (*A. johnsonii*) and 8/9 (*A. lwoffii*) have been found in specimens from general practice (Gerner-Smidt and Frederiksen, 1993; Gerner-Smidt and Tjernberg, 1993), but their biological significance is not clear. Elucidation of the ecology of the genomic species is a challenging task for the near future.

This chapter deals with metabolic properties, methods for cultivation, identification and typing of acinetobacters in clinical microbiology.

3.2 METABOLIC PROPERTIES

Bacteria belonging to the genus *Acinetobacter* are strictly aerobic, non-fermenting, oxidase-negative and catalase-positive. Bacteria belonging to this genus can utilise a wide variety of organic compounds for growth, and therefore the genus plays an important role in environmental microbiology. This explains their wide distribution in nature and probably contributes to the persistence of certain strains in the hospital environment.

Most strains are unable to reduce nitrate, which explains the previous designation *Bacterium anitratum*. Ammonium and nitrate salts can serve as sources of nitrogen (Baumann et al., 1968). The nutritional spectrum can be used for phenotypic assignment of strains to different genomic species (Bouvet and Grimont, 1986; 1987; Bouvet and Jeanjean, 1989; Gerner-Smidt et al., 1991; Gennari and Lombardi, 1993; Kämpfer et al., 1993). Strains of several genomic species produce acid from glucose, although usually they do not use this substrate for growth. The acidification of glucose is mediated by a non-specific aldolase which also mediates oxidation of lactose, D-mannose, L-arabinose and D-xylose (Baumann et al., 1968). The aldolase has been shown to be a quinoprotein with pyrroloquinoline quinone (PQQ) as a cofactor (Duine, 1991). No other carbohydrate-acidifying enzymes have been found in *Acinetobacter*.

3.3 CULTIVATION

3.3.1 Media and Growth Conditions

Acinetobacters grow readily on common, relatively simple media, including nutrient agar, trypticase soya agar and MacConkey agar. Nutrient agar with the addition of sheep or human blood is used to identify haemolysin-producing strains. For specific purposes, defined media consisting of a mineral base with one or several carbon sources have been used (Baumann et al., 1968; Juni, 1978; Bouvet and Grimont, 1986; 1987; Gerner-Smidt et al., 1991; Kämpfer et al., 1993).

Acinetobacters are strictly aerobic. Strains from clinical specimens, usually belonging to genomic species 2 (*A. baumannii*), 3 and 13TU, grow at 37°C or even at higher temperatures. In contrast, isolates of genomic species 7 (*A. johnsonii*) and 11 are unable to grow at this temperature (Gerner-Smidt et al., 1991; Kämpfer et al., 1993). A recent study by Gennari and Lombardi (1993) showed that the majority of isolates from food, belonging to genomic species 3, 5, 7, 8/9 and 10, grew at 5°C, while not all strains grew at 37°C. In order to meet the requirements of acinetobacters and other non-fermenting Gram-negative organisms, a general cultivation temperature of 30°C has been recommended (Hugh, 1978). However, considering the findings of Gennari and Lombardi (1993), a lower temperature or a combination of temperatures may be advisable, depending on the origin of the specimen.

3.3.2 Selective Growth

Selection of *Acinetobacter* strains from clinical and environmental specimens can be achieved by suppressing growth of other microorganisms in the specimens. A selective medium for acinetobacters, comprising sugars, bile salts and bromocresol purple, available commercially as Herellea Agar (Difco), has been described by Mandel et al. (1964). A modification of this medium containing vancomycin, ampicillin, cefsulodin, sugars and phenylalanine appeared to be more selective (Holton, 1983). Selection by antibiotics in more common media has been applied during several outbreaks, for example by Allen and Green (1987), who used gentamicin (10 mg/L), and by Buisson et al. (1990), who used amikacin plus vancomycin (5 mg/L). Recently, a novel antibiotic-containing medium has been described (Jawad et al., 1994). This medium, the Leeds Acinetobacter Medium (LAM), combines selectivity with differential characteristics and has been found useful for recovery of *Acinetobacter* spp. from clinical and environmental sources.

Selection of acinetobacters can also be achieved by enrichment cultivation, a methodology that has long been used by microbiologists of the "Delft School" (van Niel, 1955) and followers like Baumann (1968). Briefly, specimens containing a variety of microorganisms are applied to mineral media with a single carbon and energy source, with ammonium or nitrate salt as a nitrogen source, and a final pH of 5.5–6.0. Vigorous shaking during incubation is needed in order to outgrow pseudomonads. After 24 or 48 h, a loopful of the culture is inoculated onto solid media and, after incubation, presumptive *Acinetobacter* spp. colonies can be identified. Using this method with an acetate-enriched medium, Baumann (1968) recovered acinetobacters from 57 of 60 specimens from soil and environmental water, a finding which illustrates the wide general distribution of acinetobacters. Enrichment cultivation on acetate (see Appendix II) has also been used for the recovery of acinetobacters from faecal specimens (Grehn and von Graevenitz, 1978), and from various specimens from clinical patients and their environment (Dijkshoorn et al., 1987b).

3.4 MORPHOLOGY

Colonies of *Acinetobacter* spp. on solid media are generally smooth, sometimes mucoid, and comparable in size to those of enterobacteria (see Figure 1.3). Glucose-non-acidifying strains usually grow with smaller colonies than the glucose-acidifying groups. The colour of colonies may vary from pale yellow to greyish. Environmental strains that produce a diffusible brown pigment have been described (Pagel and Seyfried, 1976). In fluid media, e.g., nutrient broth, acinetobacters grow homogeneously with variation in turbidity. A pellicle or ring on the vessel wall at the surface of the medium can sometimes be seen, which is probably associated with the adhering properties of the strains.

In Gram-stained preparations, acinetobacters appear as Gram-negative coccoid or coccobacillary cells, or as distinct rods. Coccoid cells have a tendency to diplo arrangements, but clusters may also be found. Glucose-acidifying clinical isolates, usually belonging to genomic species 2, 3 and 13, can appear quite similar to *Neisseria* spp. because of their diplococcoid aspect. The morphology varies with the medium, growth conditions and age of the culture (Baumann et al., 1968). Gram-stain variability and variation in size and arrangement may even be observed within cultures (Baumann et al., 1968).

3.5 IDENTIFICATION METHODS

3.5.1 Identification to Genus Level

Identification of bacteria to the genus *Acinetobacter* is based on the following criteria (Cowan, 1993): Gram-negative coccobacilli, oxidase-negative (with Kovac's reagent), non-fermenting in O-F test, non-motile in hanging drop preparation, aerobic growth, catalase-positive, nitrate reduction mostly negative, and hydrolysis of Tween® mostly positive.

Identification to the genus level can also be performed by a transformation assay (Juni, 1972). In this test, DNA from any *Acinetobacter* strain is capable of transforming auxotrophs of a competent *Acinetobacter* test strain to prototrophy, which establishes a genetic relationship between the tested and competent strain. Based on this capacity, it can be concluded that the genus *Acinetobacter* is a true genomic species *sensu* Ravin (1963).

3.5.2 Identification to Genomic Species

Identification by DNA-DNA Hybridisation

The present phylogenetic definition of species is based on approximately 70% or greater DNA-DNA relatedness of strains, and phenotypic characteristics should agree with this definition. Genospecies (here called genomic species or DNA groups) which cannot be differentiated phenotypically should not be named (Wayne et al., 1987). Several hybridisation methods have been used for allocation of *Acinetobacter* isolates to genomic species, including a nitrocellulose filter method (Johnson et al., 1970), the S1 endonuclease method (Bouvet and Grimont, 1986), the hydroxyapatite method (Tjernberg and Ursing,

1989) and a quantitative bacterial dot filter method (Tjernberg et al., 1989). Of these, the dot filter method, combined with a two-step elution rather than determining $\Delta T_{m(e)}$, is probably the most simple (Tjernberg et al., 1989). Overall, these methods are laborious and only applied in special situations; further discussion of these techniques is beyond the scope of this chapter.

Phenotypic Identification by Conventional Tests

Bouvet and Grimont (1986) described a scheme of 28 phenotypic tests for identification of acinetobacters to the genomic species delineated in the same report. This system allows differentiation of 11 of 12 genomic species, leaving only two genomic species (nos. 8/9) unseparated. The identification matrix presented in this study included 255 strains, of which 74 had been allocated to genomic species by DNA-DNA hybridisation. Subsequently, Bouvet and Grimont (1987) described a simplified scheme of 16 tests which, except for tests for glucose acidification and detection of haemolysis, comprised growth temperature and carbon source utilisation tests. Gerner-Smidt et al. (1991) investigated 198 strains by this phenotypic system combined with a probability method for identification. The strains used had also been identified by DNA-DNA hybridisation (Tjernberg and Ursing, 1989). At a probability level of ≥0.95, only 78% of the strains were correctly identified, i.e., allocated to the correct genomic species. In particular, differentiation between the closely related genomic glucose-acidifying species 1, 2, 3 and 13TU (the *A. calcoaceticus-A. baumannii* complex), appeared to be difficult.

The scheme of Bouvet and Grimont (1986), combined with a computer-assisted probability method, was also used by Gennari and Lombardi (1993) to investigate 228 clinical and food isolates, but their results were not verified by DNA hybridisation. Another investigation exploring tests for the identification of genomic species was performed by Kämpfer et al. (1993). The collection of Gerner-Smidt et al. (1991) with a few additional strains, all identified by DNA-DNA hybridisation, was used. Cluster analysis was performed with 145 of the 329 criteria explored. The final proposed system comprised 22 tests, including growth at 37°C, acidification of glucose, haemolysis, a number of carbon source utilisation tests and two enzymic tests. The system allowed identification of most genomic groups, with a correct identification rate of 98% for 22 tests and 90.8% for 10 selected tests at a Willcox probability of >0.9, but genomic species 8/9 and 15TU, 4 and 6, and even 10 and 11 were difficult to separate. A summary of tests useful for phenotypic identification of most genomic species, comprising the final scheme of Kämpfer et al. (1993) and two antibiotic susceptibility tests according to

Gerner-Smidt and Frederiksen (1993), is presented in Table 3.1. Separation between the closely related genomic species 1, 2, 3 and 13TU can be obtained by the tests presented in Table 3.2. Unfortunately, no single test or even a few tests can be used for the unambiguous identification of the genomic species and, therefore, elucidation of the ecology of the different genomic species is still problematic if only phenotypic identification is used.

Phenotypic Identification by Commercial Systems

The widely used commercial identification system API 20NE comprises 21 tests, including 12 carbon source growth tests. In two studies, a total of 45 isolates from patients and the hospital environment were investigated by this system and by DNA-DNA hybridisation as a reference method (Horrevorts et al., 1995; Weernink et al., 1995). All isolates were identified correctly to the genus level by the API system. However, only five isolates (11%) were allocated correctly to the genomic species level. In the recent release of the system, API LAB Plus update V3.2.2. (01/1993), the only species in the database V5.1 are *A. baumannii*, *A. haemolyticus* and *A. lwoffii*, with *A. junii* / *A. johnsonii* as a combination, whereas the type species *A. calcoaceticus* or other genomic species are not included. In the manual of this update it is stated that *A. calcoaceticus* no longer appears in the update as the genus taxonomy was revised completely. From this it can be concluded that, at best, only a few genomic species can be identified.

The Biolog system, another commercial system, differentiates bacteria on the basis of their respiration capability with 95 different carbon sources. The reaction pattern is visualised with the redox dye tetrazolium and compared to a database. Using this system, a collection of strains from a wastewater plant were identified to the genus level with a high identification score (82.8%) (Knight et al., 1993). Identification to genomic species was, however, far from agreement with phenotypic identification by the system of Bouvet and Grimont (1987), except for genomic species 7 (*A. johnsonii*) which was highly prevalent in the collection. Another study of acinetobacters from wastewater (Soddell et al., 1993) compared Biolog with other phenotypic systems, including the systems of Bouvet and Grimont (1986; 1987) and Gerner-Smidt et al. (1991), and again discrepancies were noted. The majority of strains in this study were not identified by DNA-DNA hybridisation and, therefore, it is not possible to conclude which system was the most reliable. Bernards et al. (1995) recently explored Biolog's potential for classification and identification with 129 strains of genomic species 1-7, 8/9, 10-12 and 13-14TU, all identified by DNA-DNA hybridisation. The collection included clinical isolates from The Netherlands and Sweden, as

TABLE 3.1

Differentiation of Genomic Species of *Acinetobacter* by Phenotypic Tests.[a] Results of Genomic Species 4–17BJ are Based on Only One Strain and Should be Verified by More Authentic Strains.

	Genomic Species												
	1,2,3 and 13TU[b]	4 and 6	5	7 15TU[b]	8/9	10	11	12	13BJ 14TU	14BJ	15BJ	16BJ	17BJ
Most discriminatory tests (n = 0)													
Utilisation of													
Azelate	[+]	–	–	[–]	D	D	+	[+]	–	+	–	(+)[c]	–
β–Alanine	[+]	[–]	–	[–]	–	+	+	–	D	+	+	–	+
L–Leucine	[+]	[+]	D	[–]	–	–	–	+	[–]	+	(+)	(+)	(+)
L–Phenylalanine	[+]	–	–	–	D	–	–	+	+	+	+	+	+
L–Tryptophan	[+]	–	–	–	–	–	–	[–]	+	+	+	–	+
4–Hydroxybenzoate	[+]	[+]	–	[–]	–	D	[+]	[–]	+	+	+	–	+
Phenylacetate	[+]	–	–	–	[+]	–	D	[+]	+	+	+	+	+
Growth at													
37°C	[+]	[+]	+	–	D	+	–	+	[–]	+	+	+	+
Acid from glucose	[+]	D	–	–	D	+	–	[–]	+	+	–	–	–
Gelatinase	–	[+]	–	–	–	–	–	–	+	+	+	+	+
Additional tests (n = 4)													
Haemolysis of													
Sheep blood	–	[+]	D	–	–	–	–	–	+	+	+	+	+
Human blood	–	[+]	D	–	–	–	–	–	+	+	+	+	+
Utilisation of													
Citrate	[+]	D	D	D	[–]	+	D	–	+	+	–	+	+
Glutarate	[+]	[–]	[–]	D	D	+	+	+	D	+	–	(+)	(+)
L–Arginine	+	[+]	[+]	D	[–]	–	–	[+]	[+]	+	+	+	+
L–Aspartate	[+]	D	D	D	–	+	+	D	D	–	–	(+)	–

L–Histidine	+	+	[+]	–	–	+	+	–	+	+	+	+	+
Phenylpyruvate	[+]	–	–	–	D	–	–	[+]	+	+	+	+	+
Protocatechuate	+	+	[–]	D	–	+	+	–	+	+	+	+	+
Vanillate	[+]	[+]	–	[–]	–	D	D	–	–	+	–	+	–
Hydrolysis of[d]													
pNP–β–D–													
xylopyranoside[e]	[+]	D	–	–	–	D	–	–	–	–	–	+	+
γ–L–Glutamate–pNA[f]	[+]	D	–	–	–	–	–	–	D	–	–	+	+
Susceptibility to													
Choramphenicol	[–]	D	D	D	+	–	D	[–]	–	ND	ND	ND	ND
Penicillin G	[–]	–	D	D	[+]	–	–	D	–	ND	ND	ND	ND

[a] Adapted from Kämpfer et al. (1993), except for susceptibility tests which are according to Gerner–Smidt and Frederiksen (1993). Symbols: +, all strains positive; [+], 80% or more positive; D, 20–79% positive; [–], 20% or less positive; –, all strains negative.

[b] Genomic species grouped according to common characteristics.

[c] (+), Weak positive.

[d] Only one strain investigated.

[e] pNp, *p*–nitrophenyl; pNA, *p*–nitroanilide.

[f] Weak reaction.

TABLE 3.2

Phenotypic Differentiation of Genomic Species of the *A. calcoaceticus*–*A. baumannii* Complex[a]

	Genomic Species			
	1[b]	2[c]	3	13TU[d]
Growth at 41°C	[–]	+	d	+
Growth at 44°C	–	+	–	d
Growth with				
L–Arabinose	d	[+]	d	+
D–Glucarate	d	[+]	[+]	[–]
D–Ribose	d	[+]	d	[+]
Sorbinic acid	[–]	d	[–]	[+]
Glycerate	[+]	[–]	d	[–]
Levulinate	d	d	–	[–]
L–Tartrate	[–]	d	[+]	[–]
Acetyl–L–glutamine	[–]	d	d	d
D–Asparagine	+	–	–	[–]
L–Hydroxyproline	[–]	[+]	[–]	d

[a] Adapted from Kämpfer et al. (1993).

[b] *A. calcoaceticus.*

[c] *A. baumannii.*

[d] DNA–DNA hybridisation group 13 (Tjernberg and Ursing, 1989).

Symbols correspond to those of Table 3.1.

well as reference strains. Using Biolog database release 3.01A, 84.5% of the strains were identified to the genus *Acinetobacter*, and 45% were identified to the correct genomic group. Cluster analysis according to the breathprint pattern showed linkage between genomic species of the *A. calcoaceticus-A. baumannii* complex, between genomic species 4 (*A. haemolyticus*) and 6, and between genomic species 10 and 11. This is in agreement with the DNA homology studies of Tjernberg and Ursing (1989). Furthermore, genomic species 8/9 (*A. lwoffii*) and 12 (*A. radioresistens*) were linked together in one cluster. All strains of genomic species 5 (*A. junii*), 7 (*A. johnsonii*) and 14 were each found in separate clusters. Incorporation of these findings might improve the Biolog database for *Acinetobacter* identification.

Presumptive Identification by SDS–PAGE and Ribotyping

A study comparing cell envelope protein electrophoretic profiles and DNA hybridisation data of 120 *Acinetobacter* strains has shown that minor protein bands of 55-96 kDa can be used for differentiation between a number of DNA groups (Dijkshoorn et al., 1990a). However, this requires rigorous standardisation of electrophoretic conditions and can only be considered as a screening method.

Ribotyping of isolates of the *A. calcoaceticus-A. baumannii* complex has also shown that patterns generated by this method are DNA group-specific and can be used for species identification (Gerner-Smidt, 1992).

Identification Based on Chemical Analysis of Cell Components

Cell components are being used increasingly for classification and identification of bacteria by chemical methods like high pressure gas-liquid chromatography (HPLC), infrared spectroscopy or pyrolysis mass spectrometry. Investigation of polyamine and fatty acid composition of representative strains of all described genomic species of *Acinetobacter* showed an overall similarity in all genomic species (Kämpfer et al., 1992). Fatty acid composition was also investigated for 211 strains of *Acinetobacter* of all known genomic species as identified by DNA-DNA hybridisation (Kämpfer, 1993). Two main groups were distinguished, one with genomic species 8/9 and 15TU, and one with all other genomic species. Differentiation within the groups according to genomic species was difficult. These results suggest that fatty acid or polyamine analysis is not particularly helpful in species identification.

Molecular Methods for Species Identification

New molecular identification methods are currently being validated against DNA-DNA hybridisation. One method (Jayarao et al., 1992) is based on restriction analysis of 16S rRNA genes amplified by the polymerase chain reaction. Investigation of *Acinetobacter* strains of different genomic species by this method has indicated that a combination of patterns generated by five restriction enzymes can be used for identification of most genomic species, including genomic species 1, 2, 3 and 13TU (Vaneechoutte et al., 1995).

Another method, AFLP™, involves restriction of purified chromosomal DNA with two enzymes, ligation of adaptors, and selective amplification with specific primers, one of which is radioactively labeled (Zabeau and Vos, 1992; Huys et al., 1993; P. Janssen et al., unpublished results). Amplification products are separated on sequencing gels and analysis of patterns with approximately 50 bands is performed by the use of a computer program (GelCompar™, Applied Maths, Kortrijk, Belgium). A pilot study of 20 strains belonging to genomic species 1, 2 (*A. baumannii*), 3 and 13TU and 8 (*A. lwoffii*) has shown a species-specific grouping of patterns, while strains within the *A. calcoaceticus-A. baumannii* complex were more similar to each other than to *A. lwoffii*. Another study with approximately 150 strains of all known genomic species of *Acinetobacter* has shown a perfect correlation of AFLP™ with DNA-DNA hybridisation (P. Janssen et al., unpublished results).

A new technology is based on the use of specific rRNA probes by which microorganisms are characterised. Recently, a genus-specific 16S rRNA-targeted probe was developed to investigate the role of *Acinetobacter* spp. *in situ* (Wagner et al., 1994). Considering the rapid advances in this field, it is to be expected that species-specific probes will be available soon.

3.5.3 Conclusion

At the moment, no current method can be used for correct identification of all species and genomic species of *Acinetobacter*. In particular, identification of the clinically relevant genomic species 2 *(A. baumannii)*, 3 and 13TU by phenotypic tests is difficult (Gerner-Smidt et al., 1991), and this can be a source of confusion in clinical microbiology.

3.6 TYPING METHODS

3.6.1 Usefulness and Application of Typing

The prevalence of acinetobacters in hospitals is generally low, and these bacteria are associated more commonly with colonisation than with infection. Therefore, it does not seem appropriate to introduce typing of *Acinetobacter* into all diagnostic laboratories. During outbreaks or endemic episodes the situation is different, and a major reason for typing is then to distinguish epidemic strains from sporadics. In addition, typing methods are a tool for establishing the sources and modes of transmission, and their use may influence the strategies for prevention of spread. They can also be useful when studying the clinical significance and geographical spread of specific strains. Finally, typing results may reveal the predominance of certain types of strains which may be selected for fundamental research, for example on virulence factors.

3.6.2 General vs. Comparative Typing Methods

General or traditional typing methods, including biotyping, serotyping, phage typing and bacteriocin typing, are phenotypic methods that have long been used for the allocation of strains according to described classification schemes. In principle, the types identified in different laboratories and at different times can be compared directly, and as such the systems are universal. However, a limitation is that the systems cannot include all possible types, which implies that not all strains are typed. For example, general phage-typing schemes often have to be extended with additional phages in order to type local strains. Another limitation of serotyping, phage and bacteriocin typing systems is that the raising of antisera and collection of phages and marker strains is rather laborious and usually restricted to reference institutes.

Comparative typing ('fingerprinting') methods, including gel electrophoretic methods, are being used increasingly. Some are phenotypic, e.g., protein gel electrophoresis and extensive biotyping, but new methods are mainly genotypic. Advantages of these methods, in principle, are their applicability to different species, and that a result is obtained with each strain. Disadvantages are the relatively expensive equipment and reagents, and the required technical skills. Another disadvantage of comparative typing is that the comparison is usually restricted to a defined set of strains for several reasons: first, because comparison of electrophoretic patterns of many strains on different gels is difficult, and second, because the number of patterns can be large, which makes classification problematic. In addition, questions of how many differences in patterns are acceptable before two isolates are considered to be distinct have not yet been answered (Pitt, 1994). Computerised systems, e.g., GelCompar™ (Applied Maths, Kortrijk, Belgium) or GelManager™ (Biosystematica, Prague, Czech Republic), are being introduced to overcome these problems, but these require rigorous standardisation and multiple inclusion of control specimens. Furthermore, widescale interlaboratory comparisons and assessment of rules for classification have to be made.

In the following sections, different typing systems as applied to *Acinetobacter* will be discussed. Several of these are based on the latest taxonomic developments, while others are aimed at discrimination of strains without reference to genomic species.

3.6.3 General Typing Methods

Biotyping

A biotyping scheme comprising five tests has been used to identify 19 biotypes (Bouvet and Grimont, 1987; Bouvet et al., 1990; Table 3.3). This scheme, devised initially for typing *A. baumannii* strains, has also been used to type the closely related genomic species 3 and 13TU (Bouvet et al., 1990; Gerner-Smidt and Frederiksen, 1993; Gerner-Smidt

TABLE 3.3

Differentiation of Biotypes in Strains Belonging to the *A. calcoaceticus–A. baumannii* Complex[a,b]

	Biotype																		
Growth on	**1**	**2**	**3**	**4**	**5**	**6**	**7**	**8**	**9**	**10**	**11**	**12**	**13**	**14**	**15**	**16**	**17**	**18**	**19**
Levulinate	+	+	+	+	–	–	–	–	–	–	–	–	–	+	–	+	+	–	+
Citraconate	–	–	–	–	+	+	+	–	–	–	–	–	+	–	–	+	–	–	+
L–Phenylalanine	+	+	+	–	+	+	+	+	+	+	–	–	–	–	+	+	–	–	+
4–Hydroxybenzoate	+	+	–	–	+	+	–	+	+	–	+	–	+	+	–	+	–	+	+
L–Tartrate	+	–	–	–	+	–	–	+	–	–	–	–	–	–	+	+	+	+	–

[a] Adapted from Bouvet and Grimont (1987); Bouvet et al. (1990); Gerner-Smidt and Tjernberg (1993).

[b] Biotype 5,8,9 and 18 encountered in genomic species 3, biotype 9 and 10 in genomic species 13TU (Gerner–Smidt and Tjernberg, 1993).

and Tjernberg, 1993). The results indicated that biotypes 1, 2, 6 and 9 are associated with outbreaks in hospitals (Bouvet and Grimont, 1987; Giammanco et al., 1989; Bouvet et al., 1990; Buisson et al., 1990; Joly-Guillou et al., 1990). However, Gerner-Smidt and Tjernberg (1993) have pointed out that the majority of strains identified phenotypically to *A. baumannii* biotype 9 actually belong to genomic species 13TU, and the relatively high reported incidence of this species and biotype in outbreaks could in fact be caused by genomic species 13.

Serotyping

Various studies have suggested a large variety of surface antigens on different strains of *Acinetobacter* (Henriksen, 1973). Cross reactions between acinetobacters and streptococci or chlamydiae have been described (Heidelberger et al., 1969; Brade and Brunner, 1979). Recent studies were aimed at achieving serological differentiation between *A. baumannii* and genomic species 3 strains, i.e., the two genomic species most commonly found in hospitals. Chequer–board tube agglutination and reciprocal cross-absorptions with polyclonal rabbit immune sera against heated cells allowed identification of 34 serovars in *A. baumannii* and 26 serovars in genomic species 3 (Traub, 1989; 1990; Traub and Leonhard, 1994), but antigenic differences between *A. baumannii* and genomic species 3 serovars were not fully satisfactory. It should be noted that these DNA groups were distinguished by the 37°C growth test, which is positive for *A. baumannii* and negative for genomic species 3. Gerner-Smidt and Tjernberg (1993) have shown that strains of the unnamed genomic species 13TU and, sporadically, genomic species 3, can also be positive in this test. Therefore, reconsideration of the serotyping results in relation to unambiguous identification of the strains by DNA-DNA hybridisation could provide a more positive picture.

Phage Typing

Two complementary sets of bacteriophages have been used in several studies (Vieu et al., 1979; Santos Ferreira et al., 1984; Giammanco et al., 1989; Bouvet et al., 1990; Buisson et al., 1990; Joly-Guillou et al., 1990). The first set comprises 25 phages allowing identification of 125 phage types. The second set of 14 phages distinguishes another 25 types, here called subtypes.

Results of several studies, summarised in Table 3.4, show that 46–82% of isolates of genomic species 2 and 3 were allocated to described types or subtypes. Isolates allocated to uncommon (atypical) types varied from 3–23%, while 4–44% were unsusceptible to phages. Multiple isolates from patients (Bouvet et al., 1990) and outbreaks were included in these studies and may represent single strains. If so, the

TABLE 3.4

Phage Typing of Strains from Different Countries

Origin	Genomic Species	Allocation to: Described types and subtypes %	Allocation to: Uncommon types %	Allocation to: Insensitive %	No. of strains	Data from
Portugal	*A. calcoaceticus* sensu lato[a]	69	10	21	62	Santos Ferreira et al. (1984)
Italy	*A. baumannii* and sp.3	46	10	44	59	Giammanco et al. (1989)
The Netherlands	*A. baumannii* and sp.3	52	23	25	81	Bouvet et al. (1990)
France	*A. baumannii* and sp.3	74	3	23	103	Joly–Guillou et al. (1990)
France	*A. baumannii*	82	14	4	26	Buisson et al. (1990)

[a] Lautrop (1974).

actual percentage of untypable strains may be higher. Phage types 17, 124 and 75, and subtype 17 were commonest and were associated with epidemic spread (Giammanco et al., 1989; Bouvet et al., 1990; Buisson et al., 1990; Joly-Guillou et al., 1990).

Unfortunately, the phages are in use in one laboratory only and, to our knowledge, no comprehensive scheme for the types distinguished by the two sets of phages has been published. Although some doubt has been cast upon the reproducibility (Bouvet et al., 1990), phage typing can be useful when used in conjunction with other typing methods.

Bacteriocin Typing

Bacteriocin typing of *Acinetobacter* strains has been used in two studies. Andrews (1986) isolated 10 strains that were sensitive to bacteriocins. Using these as indicators, 176 strains were typed. Overall typability was 65%, but 56% belonged to only two groups. Stone and Das (1986) used 19 bacteriocin-containing lysates for typing 100 strains. Sixteen isolates, including 11 from an outbreak, belonged to the same type. Other types were of lower frequency or sporadic, and 54% of the strains were untypable. Bacteriocin typing of isolates identified according to the recent taxonomy has, to our knowledge, not yet been published and the question concerning the usefulness of bacteriocin typing for clinically important genomic species remains open.

3.6.4 COMPARATIVE TYPING METHODS

Biotyping

Biochemical profiles comprising large numbers of binary characteristics (test results being scored as positive or negative) or quantitative test results can be used for comparative typing of strains. Using API20NE™, Towner and Chopade (1987) distinguished 31 different biotypes in 122 strains. Patterns were stable during 1-year storage on nutrient agar slopes and it was concluded that the system can be used for monitoring the spread of *Acinetobacter* strains in hospitals. This was corroborated by Vila et al. (1989) and Horrevorts et al. (1995) who used API20NE in conjunction with other typing methods. However, the system was considered insensitive or badly reproducible in other studies (Kropec et al., 1993; M. Alexander, unpublished results).

A quantitative carbon source growth assay combined with cluster analysis showed a major grouping of isolates that was related to the epidemiological origin of the strains (Dijkshoorn et al., 1990b). Application of the method in another study allowed the discrimination of two strains of a common protein profile that were successively endemic in an intensive care unit (Dijkshoorn et al., 1993a).

The usefulness of comparative biotyping for strain identification depends on the unique features of the strains involved. Systems with large numbers of substrates, preferably commercial systems, may be of general use, but this has to be assessed with well-defined panels of strains from different outbreaks.

Antibiogram Typing

Antibiotic susceptibility patterns (antibiograms) have been used in numerous studies for comparative typing of strains and for detecting emerging resistance patterns (Bergogne-Bérézin and Joly-Guillou, 1985; Allen and Green, 1987; Alexander et al., 1988; Vila et al., 1989; Joly-Guillou et al., 1991; Struelens et al., 1993). The antibiotics used for antibiogram typing should be selected carefully, because, as pointed out in other chapters of this book, acinetobacters are inherently resistant to many agents, including β-lactam antibiotics and trimethoprim, and universally susceptible to others, e.g., polymyxins.

Patterns can be based on minimal inhibitory concentrations or on inhibition zone diameters determined by disc or tablet diffusion. Results based on inhibition zones are usually transformed to 'resistant', or 'susceptible' depending on breakpoints. This implies that useful information, i.e., the actual size of the zones, is not used. In addition, classification can be difficult if many antibiotics are used. An alternative approach that uses the full information of inhibition zones has been applied in several studies (Dijkshoorn et al., 1993b; Weernink et al., 1995; Horrevorts et al., 1995). In this method (see Appendix IV), diameters of inhibition zones are used for cluster analysis. Thus, isolates are grouped according to their similarity in the antibiogram pattern. The grouping depicted in a dendrogram can be compared to other typing characteristics and the epidemiology of the strains. Application of this method to a study of 25 isolates from five outbreaks (Dijkshoorn et al., 1993b) showed a striking grouping of isolates according to their origin except for one isolate (Figure 3.1). Furthermore, the grouping correlated well with other typing features of the isolates.

It has been emphasised that antibiogram typing results should be interpreted with caution as unrelated strains may exhibit the same antibiogram (Joly-Guillou et al., 1991). It must also be realised that changes in susceptibility may occur during endemic episodes. Nevertheless, in conjunction with other methods, antibiogram typing can be a useful tool.

Cell Envelope Protein Electrophoretic Typing

Sodium dodecyl sulphate–polyacrylamide gel electrophoresis (SDS-PAGE) of bacterial proteins has been used for discrimination of various

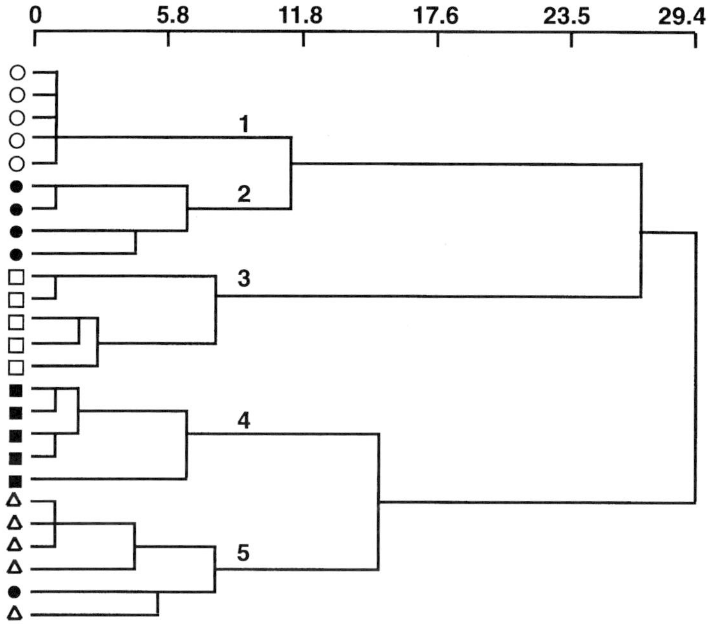

FIGURE 3.1
Dendrogram of cluster analysis based on antibiogram susceptibility patterns of isolates from five outbreaks, as determined by tablet diffusion using 12 antibiotics. Horizontal axis: dissimilarity calculated from diameters of inhibition zones using squared Euclidian distance and the unweighted pair group average linkage method. Outbreaks were in ○, Venlo, Netherlands; ●, Basildon, U.K.; □, Newcastle, U.K.; ■, The Hague, Netherlands; Δ, Odense, Denmark.

organisms (Overbeeke and Lugtenberg, 1980; Costas and Owen, 1987; Vauterin et al., 1993). The fractions analysed can be whole cell proteins or outer membranes or cell envelopes, the latter comprising the complex of layers surrounding the cytoplasm.

Cell envelope protein SDS-PAGE has been applied in a series of epidemiological and taxonomic studies. Full details of the method are given in Appendix V. Briefly, cell envelopes are obtained by sonication and fractionated centrifugation (Dijkshoorn et al., 1987a) and proteins are separated by SDS-PAGE. Rigorous standardisation of buffers and electrophoresis conditions is important for intra- and inter-gel comparison (Dijkshoorn et al., 1990a). Application of the method showed heterogeneity in unrelated strains, while multiple isolates from patients or outbreaks were indistinguishable (Dijkshoorn et al., 1987a; b). Examples of these are shown in Figure 3.2. Cell envelope protein electrophoresis has been used to trace specific strains during endemic episodes and outbreaks in hospitals (Crombach et al., 1989; Dijkshoorn et al., 1993a;

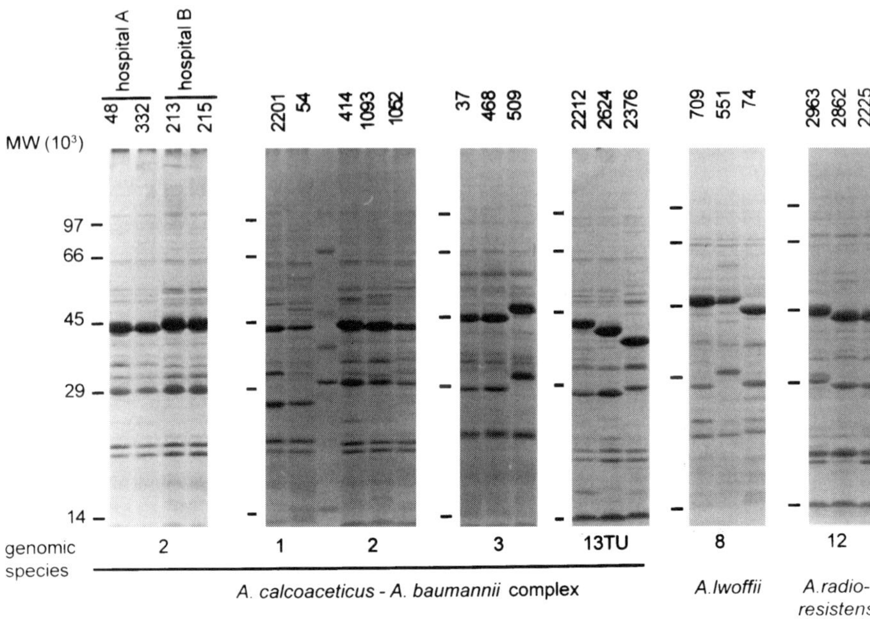

FIGURE 3.2
Cell envelope protein electrophoretic profiles of *Acinetobacter* isolates of different genomospecies, including isolates from outbreaks in two hospitals.

Horrevorts, 1995; Weernink et al., 1995). Good correlations of protein profiles, origin of strains and other typing features were found in strains from five outbreaks in different countries and cities (Dijkshoorn et al., 1993b). Interestingly, isolates from two geographically separate outbreaks appeared to be highly similar in their protein type, ribotype and biotype, and could be distinguished only by antibiogram typing. A lack of correlation of cell envelope protein patterns with epidemiological origin, biotyping and phage typing was found by Giammanco et al. (1989). Patterns in this study, however, were markedly different from those obtained by Dijkshoorn et al. (1987a; 1990a) which may be explained by differences in the gel system and staining techniques used in the two laboratories.

A taxonomic study of 120 strains (Dijkshoorn et al., 1990a) investigating cell envelope protein variation within and between genomic species showed that protein bands in the 55–100 kDa range were specific for a given DNA group, except for the closely related genomic species 1, 2, 3 and 13TU which showed common banding patterns in this range. Other bands, e.g., in the 14–20 kDa range, have been found helpful for presumptive identification of genomic species 3 and 13TU or other DNA groups (Figure 3.2). Relatedness of protein profiles and genomic species, as identified by phenotypic tests, was not clear in the study of Bouvet et al. (1990). However, this disagreement was caused by discrepancies between phenotypic identification and identification by DNA-DNA hybridisation of strains belonging to the *A. calcoaceticus-A. baumannii* complex, as can be concluded from comparison of common strains in two different studies (Bouvet et al., 1990; Dijkshoorn et al., 1990a).

Whole Cell Protein Electrophoretic Typing

Electrophoresis of whole cell protein fractions has the advantage that sample preparation is simpler than cell envelope preparation. Using this method, Mortensen et al. (1987) found similarity in *A. calcoaceticus* subsp. *anitratus* isolates from an outbreak and dissimilarity in unrelated control strains, and correlation was seen with antibiotic and plasmid typing. Correlation of whole cell protein patterns with epidemiology and antibiotic susceptibility was also found by Alexander et al. (1984; 1988) in strains of *Acinetobacter* spp. Unpublished results in our laboratory have shown that, overall, whole cell protein patterns correlate with the epidemiological origin of isolates. However, deviating whole cell protein patterns were seen in isolates from single outbreaks or from single patients, while other typing characteristics, including cell envelope protein patterns of the strains, were uniform (Figure 3.3). These findings suggest that differences between isolates from a common origin should be interpreted with caution.

Multiloci Enzyme Electrophoretic Typing

Electrophoretic investigation of relative mobilities of a large number of cellular enzymes has long been a standard method in eukaryotic population biology. The method has recently been introduced for investigating diversity in natural bacterial populations, and has been described in detail by Selander et al. (1986). Investigation of 81 *Acinetobacter* strains for 27 esterases and two dehydrogenases showed between 2 and 17 variants for each enzyme, while the pattern of grouping was in agreement with the genomic species of the strains (Picard et al., 1989). In another study of 65 clinical *Acinetobacter* strains, 14 different types were distinguished by the examination of 13 enzymes (Thurm and Ritter, 1993). One type was found in 41 multiresistant isolates with common whole cell protein profiles, and results were in agreement with the epidemiological origin of the isolates. These studies suggest that enzyme electrophoresis is a useful technique for strain identification.

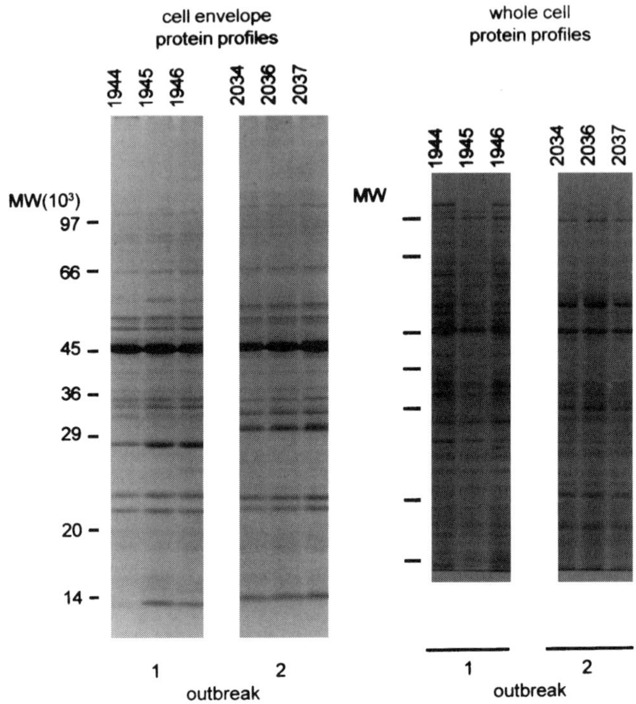

FIGURE 3.3
Comparison of cell envelope and whole cell protein profiles of six *Acinetobacter* isolates from two outbreaks.

Plasmid Typing

Plasmid typing is a relatively simple method (see Appendix IX) for strain identification and has been used for typing *Acinetobacter* strains in a number of studies. Different methods have been used, some derived from the method of Birnboim and Doly (1979) (e.g., Getchell-White et al., 1989; Johnson et al., 1992; Kropec et al., 1993). Other methods have been described in detail by Alexander et al. (1988), Gerner–Smidt (1989) and Seifert et al. (1994a). Strains with similar plasmid patterns can be differentiated further by restriction endonuclease digestion of plasmid DNA followed by electrophoresis (Vila et al., 1989; Kropec et al., 1993), or by hybridisation of plasmids with a labelled probe from one of the strains (Gerner-Smidt and Tjernberg, 1993).

With few exceptions, plasmid typing was combined in these studies with other typing methods, including antibiogram typing, biotyping, protein profiling or ribotyping. Identification according to the latest taxonomy was performed by the phenotypic system of Bouvet and Grimont (1986) in two studies (Seifert et al., 1994a; 1994b), and in one study by DNA-DNA hybridisation (Gerner-Smidt and Tjernberg, 1993). In other studies, identification was based on the old taxonomy or was performed with commercial systems that do not identify all recent species.

Plasmids in strains were variable in number and size. For example, Gerner-Smidt and Tjernberg (1993) found no plasmids in 10/23 strains and between one and four plasmids of 2.1 to >100 kb in 13 other strains of the *A. calcoaceticus-A. baumannii* complex. Seifert et al. (1994a) found plasmids in 84.4% of *Acinetobacter* sp. 3 strains, and in all isolates of *A. junii, A. johnsonii* and *A. lwoffii*. Relatively low numbers of plasmids in glucose-acidifying strains, most of which belong to the *A. calcoaceticus-A. baumannii* complex, were also observed in other studies (Gerner-Smidt, 1989; Kropec et al., 1993).

Correlation of plasmid content with antibiotic susceptibility and/or biotype was observed in several studies (Getchell-White et al., 1989; Vila et al., 1989), but lack of agreement with antibiograms or protein profiles has also been observed (Alexander et al., 1988; Johnson et al., 1992).

Analysis of plasmid profiles has been useful for delineating several outbreaks (Hartstein et al., 1988; Vila et al., 1989; Beck-Sagué et al., 1990; Kropec et al., 1993). Seifert et al. (1994a) used a well-defined collection of strains of genomic species 3, *A. junii, A. johnsonii* and *A. lwoffii*, and showed related patterns in multiple isolates from patients. Isolates from different patients were clearly distinct.

Based on these studies it appears that plasmid typing can be useful in epidemiological studies, provided that results are interpreted in relation to other typing methods and the origin of the strains is well described.

Ribotyping

Characterisation of microorganisms by ribotyping has appeared useful in a number of reports (Owen, 1989; Grimont and Grimont, 1986). The method as used for *Acinetobacter* has been described in detail by Gerner-Smidt (1992). Briefly, it comprises purification of DNA, digestion by restriction enzymes, separation of resulting fragments by agarose gel electrophoresis, and transfer to nylon membrane by vacuoblotting. Selective banding patterns are visualised by hybridisation with a non-radioactive digoxigenin-11-dUTP-labelled probe of cDNA from rRNA of *E. coli* (see Appendix VIII). By this method Gerner-Smidt (1992) investigated 70 strains that were identified as either *A. calcoaceticus, A. baumannii* or the unnamed genomic species 3 and 13TU by DNA-DNA hybridisation. Patterns generated by restriction with *Eco*RI, *Cla*I or *Sal*I were specific for each genomic group and are thus useful for

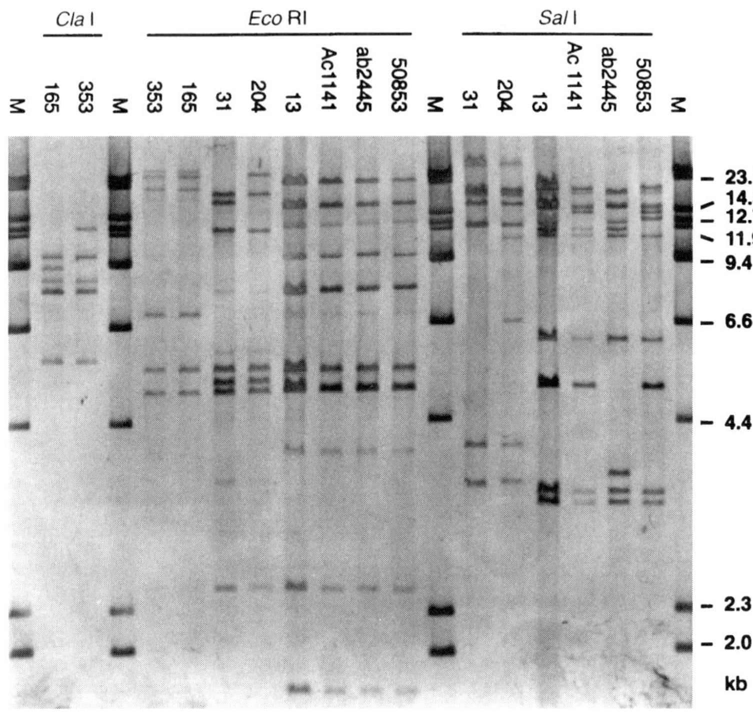

FIGURE 3.4
Ribotypes of clinical *Acinetobacter* species. Eight isolates with similar *Eco*RI ribotypes but different *Sal*I or *Cla*I ribotypes are shown. The size markers are phage λ cut with *Hin*dIII. (Courtesy of P. Gerner-Smidt.)

identification. Reproducibility was excellent, and the discriminatory power for typing by combined use of the three enzymes was high, with 52 different patterns in 70 unrelated strains.

Another ribotyping study of strains of *A. baumannii* and the unnamed groups 3 and 13TU showed that patterns within outbreaks were stable and correlated with results of other typing methods (Dijkshoorn et al., 1993b). The usefulness of ribotyping for studying the epidemiology of *Acinetobacter* is demonstrated clearly in Figure 3.4 (courtesy of Dr. P. Gerner-Smidt) showing the diversity of patterns generated by three enzymes, and by Figure 3.5 showing ribotypes that demonstrate the epidemic spread of a strain following transfer of a patient from one hospital to another.

Although the diversity of ribotypes within the *A. calcoaceticus*-*A. baumannii* complex is considerable, common patterns in apparently unrelated strains have been found (Gerner-Smidt, 1992; Dijkshoorn et

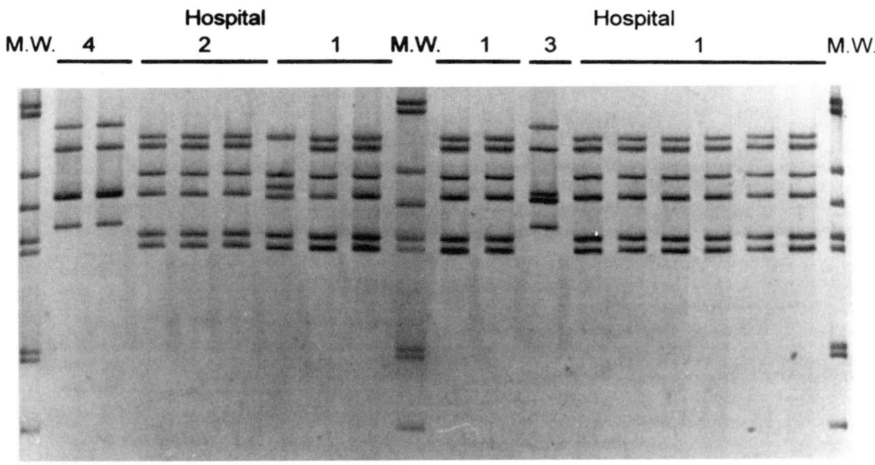

FIGURE 3.5
*Cla*I ribotypes of Danish *Acinetobacter* isolates obtained from patients in four hospitals. An epidemic caused by a single strain of *A. baumannii* (DNA group 2) has spread from a burns unit in hospital 1 to hospital 2 with the transfer of one patient from the burns unit to hospital 2. The profile of one isolate (the first from hospital 1) is typical for DNA group 2, but clearly different from that of the epidemic strain. This isolate is unrelated to the epidemic. The isolates from hospitals 3 and 4 are also related to the epidemic strain and show typical DNA group 3 patterns. The size markers are a mixture of phage λ cut with *Hin*dIII and *Sty*I. (Courtesy of P. Gerner-Smidt.)

al., 1993b), which may indicate a clonal relationship. This is corroborated by an unpublished multinational study investigating *A. baumannii* isolates from outbreaks and sporadic isolates. A number of apparently unrelated outbreak strains had identical patterns, while other strains were heterogeneous.

In summary, several studies have shown that ribotyping, although laborious, is highly useful for investigating outbreaks and for studying questions of general epidemiological importance, and also for species identification.

Pulsed-Field Gel Electrophoresis

Analysis of restriction fragment length polymorphisms by pulsed–field gel electrophoresis (PFGE) is becoming a standard method in hospital epidemiology. In this method, genomic DNA is cut infrequently with enzymes that generate large fragments, which are then separated by special electrophoretic conditions (Birren and Lai, 1993; see Appendix VI). The method has been applied in recent studies of *Acinetobacter* comparing patterns obtained by restriction enzymes *Apa*I

(Gräser et al., 1993; Seifert et al., 1994b, Tankovic et al., 1994), *Sma*I (Allardet–Servent et al., 1989), *Apa*I and *Sma*I (Gouby et al., 1992), *Apa*I and *Sfi*I (L. Dijkshoorn and M. Geneugelijk, unpublished results), and *Nhe*I and *Sma*I (Struelens et al., 1993). Results indicated considerable DNA polymorphism in *A. baumannii*, even within biotypes, and overall uniformity in outbreaks or multiple isolates from single patients. Longitudinal reproducibility was good (Seifert et al., 1994b) and results were in agreement with those obtained by plasmid typing and PCR fingerprinting (Gräser et al., 1993; Struelens et al., 1993; Seifert et al., 1994b). Comparison of PFGE with ribotyping for strains of the *A. calcoaceticus-A. baumannii* complex showed that PFGE was more discriminatory, but patterns generated by the latter method were not species specific (H. Seifert and P. Gerner-Smidt, unpublished results).

Equipment and some of the enzymes used for PFGE are costly, while *in situ* DNA preparation, enzymic digestion and electrophoresis require several days. Nevertheless, PFGE results provide highly useful information in conjunction with other typing methods.

Polymerase Chain Reaction (PCR) Fingerprinting

Fingerprinting on the basis of amplification patterns of variable DNA sequences by PCR (see Appendix VII) is being used increasingly for typing microorganisms (Maslow et al., 1993; van Belkum, 1994). The technology has been applied to delineating the relatedness of *A. baumannii* strains by Gräser et al. (1993) who used the core region of phage M13 as a single primer. Isolates from single outbreaks were uniform in PCR fingerprints and PFGE patterns. Struelens et al. (1993) used ERIC1 and ERIC2 (Versalovic et al., 1991) as single primers for typing *A. baumannii* isolates from clinical specimens in a tertiary care hospital. General agreement was observed between PCR fingerprinting, macrorestriction analysis by PFGE and the epidemiological origin of the strains. In another study (Reboli et al., 1994), a combination of ERIC1R and ERIC2 primers was not useful, whereas REP1-RI plus REP2-I allowed the relatedness between strains to be determined. L. Dijkshoorn et al. (unpublished results) have investigated PCR with primer combinations ERIC1/ERIC2 and ERIC2/1026 (van Belkum et al., 1993) and a collection of strains identified as belonging to the *A. calcoaceticus-A. baumannii* complex by DNA-DNA hybridisation. A great variety of patterns was observed in unrelated strains, while common patterns were observed in multiple isolates from single patients or outbreaks. Results were in agreement with cell envelope protein pattern and biotyping results obtained previously.

Slight variations in the intensity of banding patterns can occur in isolates from common origin (Gräser et al., 1993; L. Dijkshoorn et al.,

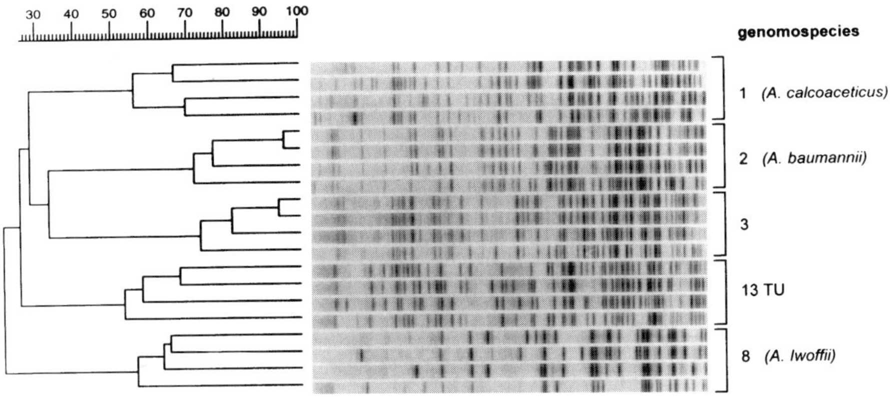

FIGURE 3.6
Amplified fragment length polymorphism (AFLP™) patterns of strains belonging to genomic species 1, 2 (*A. baumannii*), 3 and 13TU and 8 (*A. lwoffii*). (Courtesy of P. Janssen.)

unpublished results), and further studies are required to assess the longitudinal reproducibility. Nevertheless, PCR seems a useful method for simple and rapid typing of acinetobacters, especially for tracing epidemic strains during outbreaks, although combination with other typing methods is advocated.

High-resolution fingerprints can be obtained by AFLP™, which is an elaborate PCR application resulting in patterns comprising approximately 50 bands on sequencing gels (see Section 3.5.2). This method is not expected to be used for rapid strain identification, but is more suited for studying fundamental questions concerning the clonal relatedness of strains (Figure 3.6).

3.7 CONCLUSIONS

Phenotypic studies of large numbers of strains validated by DNA homology studies have shown that some genomic groups of *Acinetobacter* can only be identified unambiguously by DNA-DNA hybridisation. In particular, genomic species of the *A. calcoaceticus-A. baumannii* complex, which are clinically the most relevant DNA groups, are difficult to separate. Therefore, as stated by Weaver and Actis (1994), authors should indicate how their *Acinetobacter* species were identified, and identification should be considered and stated to be presumptive unless an extensive set of assimilation tests or DNA hybridisation tests were used. It is to be expected that additional molecular methods which may contribute to correct species identification will soon be developed.

The high degree of relatedness of some genomic species at different levels of genetic expression has led to suggestions for unification of these into species that can be identified unambiguously (Tjernberg and Ursing, 1989; Dijkshoorn et al., 1990a; Kämpfer et al., 1993). However, new changes in the taxonomy will also contribute to the already existing confusion. For the time being, a proposal has been made to combine genomic species 1, 2, 3 and 13 into the *A. calcoaceticus-A. baumannii* complex (Gerner-Smidt et al., 1991).

For epidemiological typing, a range of methods is now available to discriminate *Acinetobacter* strains, with or without any reference to genomic species (DNA group). Several studies have indicated that differences in acinetobacters from clinical specimens and outbreaks can be small. Therefore, combinations of methods, including at least one molecular method, are recommended before designating apparently indistinguishable isolates as the same strain. This strategy has been followed increasingly in recent years and typing of *Acinetobacter* has evolved considerably as a result.

Finally, since a variety of methods are becoming more current, standardisation of procedures should be emphasised in order to allow longitudinal and interlaboratory comparison, and to allow questions concerning the geographic spread and pathogenicity of particular strains to be addressed.

REFERENCES

Alexander, M., Ismail, F., Jackman, P.J.H. and Noble, W.C. (1984). Fingerprinting *Acinetobacter* strains from clinical sources by numerical analysis of electrophoretic protein patterns. *J. Med. Microbiol.* 18, 55–64.

Alexander, M., Rahman, M., Taylor, M. and Noble, W.C. (1988). A study of the value of electrophoretic and other techniques for typing *Acinetobacter calcoaceticus*. *J. Hosp. Infect.* 12, 273–287.

Allardet-Servent, A., Bouziges, N., Carles-Nurit, M.J., Bourg, G., Gouby, A. and Ramuz, M. (1989). Use of low-frequency-cleavage restriction endonucleases for DNA analysis in epidemiological investigations of nosocomial bacterial infections. *J. Clin. Microbiol.* 27, 2057–2061.

Allen, K.D. and Green, H.T. (1987). Hospital outbreak of multi-resistant *Acinetobacter anitratus*: an airborne mode of spread? *J. Hosp. Infect.* 9, 110–119.

Andrews, H.J. (1986). Acinetobacter bacteriocin typing. *J. Hosp. Infect.* 7, 169–175.

Baumann, P. (1968). Isolation of *Acinetobacter* from soil and water. *J. Bacteriol.* 96, 39–42.

Baumann, P., Doudoroff, M. and Stanier, R.Y. (1968). A study of the *Moraxella* group II. Oxidative–negative species (genus *Acinetobacter*). *J. Bacteriol.* 95, 1520–1541.

Beck-Sagué, C.M., Jarvis, W.R., Brook, J.H., Culver, D.H., Potts, A., Gay, E., Shotts, B.W., Hill, B., Anderson, R.L. and Weinstein, M.P. (1990). Epidemic bacteremia due to *Acinetobacter baumannii* in five intensive care units. *Am. J. Epidemiol.* 132, 723–733.

Beijerinck, M.W. (1911). Über Pigmentbildung bei Essigbakterien. *Centralbl. Bakteriol. Parasit.* 29, 169–176.

Bergogne-Bérézin, E. and Joly-Guillou, M.-L. (1985). An underestimated nosocomial pathogen, *Acinetobacter calcoaceticus*. *J. Antimicrob. Chemother.* 16, 535–538.

Bernards, A.T., Dijkshoorn, L., van der Toorn, J., Bochner, B.R. and van Boven, C.P.A. (1995). Phenotypic characterisation of *Acinetobacter* strains of 13 DNA-DNA hybridisation groups by means of the Biolog system. *J. Med. Microbiol.* 42, 113–119.

Birnboim, H.C. and Doly, J. (1979). A rapid alkaline extraction procedure for screening recombinant plasmid DNA. *Nucl. Acids Res.* 7, 1513–1523.

Birren, B. and Lai, E. (1993). *Pulsed Field Gel Electrophoresis: A Practical Guide.* Academic Press, London.

Bouvet, P.J.M. and Grimont, P.A.D. (1986). Taxonomy of the genus *Acinetobacter* with the recognition of *Acinetobacter baumannii* sp. nov., *Acinetobacter haemolyticus* sp. nov., *Acinetobacter johnsonii* sp. nov., and *Acinetobacter junii* sp. nov. and emended descriptions of *Acinetobacter calcoaceticus* and *Acinetobacter lwoffii*. *Int. J. Syst Bacteriol.* 36, 228–240.

Bouvet, P.J.M. and Grimont, P.A.D. (1987). Identification and biotyping of clinical isolates of *Acinetobacter*. *Ann. Inst. Pasteur Microbiol.* 138, 569–578.

Bouvet, P.J.M. and Jeanjean, S. (1989). Delineation of new proteolytic genomic species in the genus *Acinetobacter*. *Res. Microbiol.* 140, 291–299.

Bouvet, P.J.M., Jeanjean, S., Vieu, J.F. and Dijkshoorn L. (1990). Species, biotype, and bacteriophage type determinations compared with cell envelope protein profiles for typing *Acinetobacter* strains. *J. Clin. Microbiol.* 28, 170–176.

Brade, H. and Brunner, H. (1979). Serological cross-reactions between *Acinetobacter calcoceticus* and chlamydiae. *J. Clin. Microbiol.* 10, 819–822.

Buisson, Y., Tran Van Nhieu, G., Ginot, L., Bouvet, P., Schill, H., Driot, L. and Meyran, M. (1990). Nosocomial outbreaks due to amikacin-resistant tobramycin-sensitive *Acinetobacter* species: correlation with amikacin usage. *J. Hosp. Infect.* 15, 83–93.

Costas, M. and Owen, R.J. (1987). Numerical analysis of electrophoretic protein patterns of *Streptobacillus moniliformis* strains from human, murine and avian infections. *J. Med. Microbiol.* 23, 303–311.

Cowan, S.T. (1993). *Cowan and Steel's Manual for the Identification of Medical Bacteria*, 3rd ed. G.I. Barrow and R.K.A. Feltham (Eds). University Press, Cambridge.

Crombach, W.H.J., Dijkshoorn, L., van Noort-Klaassen, M., Niessen, J. and van Knippenberg-Gordebeke, G. (1989). Control of an epidemic spread of a multi-resistant strain of *Acinetobacter calcoaceticus* in a hospital. *Intensive Care Med.* 15, 166–170.

Dijkshoorn, L., Michel M.F. and Degener, J.E. (1987a). Cell envelope protein profiles of *Acinetobacter calcoaceticus* strains isolated in hospitals. *J. Med. Microbiol.* 23, 313–319.

Dijkshoorn, L., van Vianen, W., Degener, J. E. and Michel M.F. (1987b). Typing of *Acinetobacter calcoaceticus* strains isolated from hospital patients by cell envelope protein profiles. *Epidemiol. Infect.* 99, 659–667.

Dijkshoorn, L., Tjernberg, I., Pot, B., Michel M.F., Ursing, J. and Kersters, K. (1990a). Numerical analysis of cell envelope protein profiles of *Acinetobacter* strains classified by DNA-DNA hybridization. *Syst. Appl. Microbiol.* 13, 338–344.

Dijkshoorn, L., van Ooyen, A., Hop, W.C.J., Theuns M. and Michel, M.F. (1990b). Comparison of clinical acinetobacter strains using a carbon source growth assay. *Epidemiol. Infect.* 104, 443–453.

Dijkshoorn, L., van Dalen, R., van Ooyen, A., Bijl, D., Tjernberg, I., Michel, M.F. and Horrevorts, A.M. (1993a). Endemic acinetobacter in intensive care units: Epidemiology and clinical impact. *J. Clin. Pathol.* 46, 533–536.

Dijkshoorn, L., Aucken, H.M., Gerner-Smidt, P., Kaufmann, M.E., Ursing, J. and Pitt, T.L. (1993b). Correlation of typing methods for *Acinetobacter* isolates from hospital outbreaks. *J. Clin. Microbiol.* 31, 702–705.

Duine, J.A. (1991). Energy generation and the glucose dehydrogenase pathway in *Acinetobacter*. In *The Biology of* Acinetobacter, pp. 295–312. K.J. Towner, E. Bergogne-Bérézin and C.A. Fewson (Eds). Plenum Press, New York.

Gennari, M. and Lombardi, P. (1993). Comparative characterization of *Acinetobacter* strains isolated from different foods and clinical sources. *Zentralbl. Bakteriol.* 279, 553–564.

Gerner-Smidt, P. (1989). Frequency of plasmids in strains of *Acinetobacter calcoaceticus*. *J. Hosp. Infect.* 14, 23–28.

Gerner-Smidt, P. (1992). Ribotyping of the *Acinetobacter calcoaceticus-Acinetobacter baumannii* complex. *J. Clin. Microbiol.* 30, 2680–2685.

Gerner-Smidt, P., Tjernberg, I. and Ursing, J. (1991). Reliability of phenotypic tests for identification of *Acinetobacter* species. *J. Clin. Microbiol.* 29, 277–282.

Gerner-Smidt, P. and Frederiksen, W. (1993). *Acinetobacter* in Denmark. I. Taxonomy, antibiotic susceptibility, and pathogenicity of 112 clinical strains. *APMIS* 101, 815–825.

Gerner-Smidt, P. and Tjernberg, I. (1993). *Acinetobacter* in Denmark. II. Molecular studies of the *Acinetobacter calcoaceticus-Acinetobacter baumannii* complex. *APMIS* 101, 826–832.

Getchell-White, S.I., Donowitz, L.G. and Gröschel, D.H.M. (1989). The inanimate environment of an intensive care unit as a potential source of nosocomial bacteria. Evidence for long survival of *Acinetobacter calcoaceticus*. *Infect. Control Hosp. Epidemiol.* 10, 402–407.

Giammanco, A., Vieu, J.F., Bouvet, P.J.M., Sarzana, A. and Sinatra, A. (1989). A comparative assay for epidemiological markers for *Acinetobacter* strains isolated in a hospital. *Zentralbl. Bakteriol.* 272, 231–241.

Gouby, A., Carles-Nurit, M.J., Bouziges, N., Bourg, G., Mesnard R. and Bouvet, P.J.M. (1992). Use of pulsed-field gel electrophoresis for investigation of hospital outbreaks of *Acinetobacter baumannii*. *J. Clin. Microbiol.* 30, 1588–1591.

Gräser, Y., Klare, I., Halle, E., Gantenberg, R., Buchholz, P., Jacobi, H.D., Presber, W. and Schönian, G. (1993). Epidemiological study of an *Acinetobacter baumannii* outbreak by using polymerase chain reaction fingerprinting. *J. Clin. Microbiol.* 31, 2417–2420.

Grehn, M. and von Graevenitz, A. (1978). Search for *Acinetobacter calcoaceticus* subsp. *anitratus:* enrichment of faecal samples. *J. Clin. Microbiol.* 8, 342–343.

Grimont, F. and Grimont, P. (1986). Ribosomal ribonucleic acid gene restriction patterns as potential taxonomic tools. *Ann. Inst. Pasteur Microbiol.* 137B, 165–175.

Hartstein, A.I., Rashad, A.L., Liebler, J.M., Actis, L.A., Freeman, J., Rourke, J.W., Stibolt, T.B., Tolmasky, M.E., Ellis, G.R. and Crosa, J.H. (1988). Multiple intensive care unit outbreak of *Acinetobacter calcoaceticus* subspecies *anitratus* respiratory infection and colonization associated with contaminated, reusable ventilator circuits and resuscitation bags. *Am. J. Med.* 85, 624–631.

Heidelberger, M., Das, A. and Juni, E. (1969). Immunochemistry of the capsular polysaccharide of an *Acinetobacter*. *Proc. Natl. Acad. Sci. U.S.* 63, 47–50.

Henriksen, S.D. (1973). *Moraxella*, *Acinetobacter*, and the *Mimeae*. *Bacteriol. Rev.* 37, 522–561.

Holton, J. (1983). A note on the preparation and use of a selective differential medium for the isolation of *Acinetobacter* spp. from clinical sources. *J. Appl. Bacteriol.* 54, 141–142.

Horrevorts, A.M., Bergman, K., Kolleé, L., Breuker, I., Tjernberg, I. and Dijkshoorn, L. (1995). Clinical and epidemiological investigations of *Acinetobacter* genomospecies 3 in a neonatal intensive care unit. *J. Clin. Microbiol.* 33, 1567–1572.

Hugh, R. (1978). Classical methods for isolation and identification of glucose nonfermenting Gram-negative rods. In *Glucose Nonfermenting Gram-Negative Bacteria in Clinical Microbiology*, p. 2. G.L. Gilardi (Ed.). CRC Press, Boca Raton, FL.

Huys, G., Coopman, R., Vancanneyt, M., Kersters, I., Verstraete, W., Kersters, K. and Janssen, P. (1993). High resolution differentiation of aeromonads. *Med. Microbiol. Lett.* 2, 248–255.

Jawad, A., Hawkey, P.M., Heritage, J. and Snelling, A.M. (1994). Description of Leeds Acinetobacter Medium, a new selective and differential medium for isolation of clinically important *Acinetobacter* spp., and comparison with Herellea agar and Holton's agar. *J. Clin. Microbiol.* 32, 2353–2358.

Jayarao, B.M., Doré, J.J.E. and Oliver, S.P. (1992). Restriction fragment length polymorphism analysis of 16S ribosomal DNA of *Streptococcus* and *Enterococcus* species of bovine origin. *J. Clin. Microbiol.* 30, 2235–2240.

Johnson, J.L., Anderson, R.S. and Ordal, E.J. (1970). Nucleic acid homologies among oxidase-negative *Moraxella* species. *J. Bacteriol.* 101, 568–573.

Johnson, D.R., Love-Dixon, M.A., Brown, W.J., Levine, D.P., Downes, F.P. and Hall, W.N. (1992). Delayed detection of an increase in resistant *Acinetobacter* at a Detroit hospital. *Infect. Control Hosp. Epidemiol.* 13, 394–398.

Joly-Guillou, M.-L., Bergogne-Bérézin, E. and Vieu, J.F. (1990). A study of the relationships between antibiotic resistance phenotypes, phage-typing and biotyping of 117 clinical isolates of *Acinetobacter* spp. *J. Hosp. Infect.* 16, 49–58.

Joly-Guillou, M.-L., Bergogne-Bérézin, E. and Vieu, J.F. (1991). Epidemiology of *Acinetobacter* strains isolated from nosocomial infections in France. In *The Biology of* Acinetobacter, pp. 63–82. K.J. Towner, E. Bergogne-Bérézin and C.A. Fewson (Eds). Plenum Press, New York.

Juni, E. (1972). Interspecies transformation of *Acinetobacter*: genetic evidence for a ubiquitous genus. *J. Bacteriol.* 112, 917–931.

Juni, E. (1978). Genetics and physiology of *Acinetobacter*. *Annu. Rev. Microbiol.* 32, 349–371.

Kämpfer, P. (1993). Grouping of *Acinetobacter* genomic species by cellular fatty acid composition. *Med. Microbiol. Lett.* 2, 394–400.

Kämpfer, P., Bark, K., Busse, H.J., Auling, G. and Dott, W. (1992). Numerical and chemotaxonomy of polyphosphate accumulating *Acinetobacter* strains with high polyphosphate:AMP phosphotransferase (PPAT) activity. *Syst. Appl. Microbiol.* 15, 409–419.

Kämpfer, P., Tjernberg, I. and Ursing, J. (1993). Numerical classification and identification of *Acinetobacter* genomic species. *J. Appl. Bacteriol.* 75, 259–268.

Knight, G.C., McDonnell, S.A., Seviour, R.J. and Soddell, J.A. (1993). Identification of *Acinetobacter* isolates using the Biolog identification system. *Lett. Appl. Microbiol.* 16, 261–264.

Kropec, A., Hübner, J. and Daschner, F.D. (1993). Comparison of three typing methods in hospital outbreaks of *Acinetobacter calcoaceticus* infection. *J. Hosp. Infect.* 23, 133–141.

Lautrop, H. (1974). *Acinetobacter*. In *Bergey's Manual of Determinative Bacteriology*, 8th ed., pp. 436–438. R.E. Buchanan and N.E. Gibbons (Eds). Williams & Wilkins, Baltimore.

Mandel, A.D., Wright, K. and McKinnon, J.M. (1964). Selective medium for isolation of *Mima* and *Herellea* organisms. *J. Bacteriol.* 88, 1524–1525.

Maslow, J.N., Mulligan, M.E. and Arbeit, R.D. (1993). Molecular epidemiology: application of contemporary techniques to the typing of microorganisms. *Clin. Infect. Dis.* 17, 153–164.

Mortensen, J.E., LaRocco, M.T., Steiner, B. and Ribner, B. (1987). Protein fingerprinting for the determination of relatedness in *Acinetobacter calcoaceticus* subspecies *anitratus* isolated from patients in a surgical intensive care unit. *Infect. Control* 8, 512–515.

Nishimura, Y., Ino, T. and Iizuka, H. (1988) *Acinetobacter radioresistens* sp. nov. isolated from cotton and soil. *Int. J. Syst. Bacteriol.* 38, 209–211.

Overbeeke, N. and Lugtenberg, B. (1980). Major outer membrane proteins of *Escherichia coli* strains of human origin. *J. Gen. Microbiol.* 121, 373–380.

Owen, R.J. (1989). Chromosomal DNA fingerprinting — a new method of species and strain identification applicable to microbial pathogens. *J. Med. Microbiol.* 30, 89–99.

Pagel, J.E. and Seyfried, P.L. (1976). Numerical taxonomy of aquatic acinetobacter isolates. *J. Gen. Microbiol.* 95, 220–232.

Picard, B., Goullet, P., Bouvet, P.J.M., Decoux, G. and Denis, J.B. (1989). Characterization of bacterial genomic species by computer-assisted statistical analysis of enzyme electrophoresis data. *Electrophoresis.* 10, 680–685.

Pitt, T.L. (1994). Bacterial typing systems: the way ahead. *J. Med. Microbiol*, 40, 1–2.

Ravin, A.W. (1963). Experimental approaches to the study of bacterial phylogeny. *Am. Nat.* 97, 307–318.

Reboli, A.C., Houston, E.D., Monteforte, J.S., Wood, C.A. and Hamill, R.J. (1994). Discrimination of epidemic and sporadic isolates of *Acinetobacter baumannii* by repetitive element PCR-mediated DNA fingerprinting. *J. Clin. Microbiol*, 32, 2635–2640.

Santos Ferreira, M.O., Vieu, J.F. and Klein, B. (1984). Phage-types and susceptibility to 26 antibiotics of nosocomial strains of *Acinetobacter* isolated from Portugal. *J. Int. Med. Res.* 12, 364–368.

Seifert, H., Schulze, A., Baginsky, R. and Pulverer, G. (1994a). Plasmid DNA fingerprinting of *Acinetobacter* species other than *Acinetobacter baumannii*. *J. Clin. Microbiol.* 32, 82–86.

Seifert, H., Schulze, A., Baginsky, R. and Pulverer, G. (1994b). Comparison of four different typing methods for epidemiologic typing of *Acinetobacter baumannii*. *J. Clin. Microbiol.* 32, 1816–1819.

Selander, R.K., Caugant, D.A., Ochman, H., Musser, J.M., Gilmour, M.N. and Whittam, T.S. (1986). Methods of multilocus enzyme electrophoresis for bacterial population genetics and systematics. *Appl. Environ. Microbiol.* 51, 873–884.

Soddell, J.A., Beachman, A.M. and Seviour, R.J. (1993). Phenotypic identification of non-clinical isolates of *Acinetobacter* species. *J. Appl. Bacteriol.* 74, 210–214.

Stone, J.W. and Das, B.C. (1986). Investigation of an outbreak of infection with *Acinetobacter calcoaceticus* in a special care baby unit. *J. Hosp. Infect.* 7, 42–48.

Struelens, M. J., Carlier, E., Maes, N., Serruys, E., Quint, W.G.V. and van Belkum, A. (1993). Nosocomial colonization and infection with multiresistant *Acinetobacter baumannii*: outbreak delineation using DNA macrorestriction analysis and PCR-fingerprinting. *J. Hosp. Infect.* 25, 15–32.

Tankovic, J., Legrand, P., De Gatines, G., Chemineau, V., Brun-Buisson, C. and Duval, J. (1994). Characterization of a hospital outbreak of imipenem-resistant *Acinetobacter baumannii* by phenotypic and genotypic methods. *J. Clin. Microbiol.* 32, 2677–2681.

Thurm, V. and Ritter, E. (1993). Genetic diversity and clonal relationships of *Acinetobacter baumannii* strains isolated in a neonatal ward: epidemiological investigations by allozyme, whole-cell protein and antibiotic resistance analysis. *Epidemiol. Infect.* 111, 491–498.

Tjernberg, I. and Ursing, J. (1989). Clinical strains of *Acinetobacter* classified by DNA-DNA hybridization. *APMIS* 97, 595–605.

Tjernberg, I., Lindh, E. and Ursing, J. (1989). A quantitative bacterial dot method for DNA-DNA hybridization and its correlation to the hydroxyapatite method. *Curr. Microbiol.* 18, 77–81.

Towner, K.J. and Chopade, B.A. (1987). Biotyping of *Acinetobacter calcoaceticus* using the API20NE system. *J. Hosp. Infect.* 10, 145–151.

Traub, W.H. (1989). *Acinetobacter baumannii* serotyping for delineation of outbreaks of nosocomial cross-infection. *J. Clin. Microbiol.* 27, 2713–2716.

Traub, W.H. (1990). Serotyping of clinical isolates of acinetobacter: serovars of genospecies 3. *Zentralbl. Bakteriol.* 273, 12–23.

Traub, W.H. and Leonhard, B. (1994). Serotyping of *Acinetobacter baumannii* and genospecies 3: an update. *Med. Microbiol. Lett.* 3, 120–127.

van Belkum, A. (1994). DNA fingerprinting of medically important microorganisms by use of PCR. *Clin. Microbiol. Rev.* 7, 174–184.

van Belkum, A., Bax, R., Peerbooms, P., Goessens, W.H.F., van Leeuwen, N. and Quint, W.G.V. (1993). Comparison of phage typing and DNA fingerprinting by polymerase chain reaction for discrimination of methicillin-resistant *Staphylococcus aureus* strains. *J. Clin. Microbiol.* 31, 798–803.

Vaneechoutte, M., Dijkshoorn, L., Tjernberg, I., Elaichouni, A., De Vos, P., Claeys, G. and Verschraegen, G. (1995). Identification of *Acinetobacter* genomic species by amplified ribosomal DNA restriction analysis. *J. Clin. Microbiol.* 33, 11–15.

van Niel, C.B. (1955). Natural selection in the microbial world. *J. Gen. Microbiol.* 13, 201–217.

Vauterin, L., Swings, J. and Kersters, K. (1993). Protein electrophoresis and classification. In *Handbook of New Bacterial Systematics*, pp. 251–280. M. Goodfellow and A.G. O'Donnell (Eds). Academic Press, London.

Versalovic, J., Koeuth, T. and Lupski, J.R. (1991). Distribution of repetitive DNA sequences in eubacteria and application to fingerprinting of bacterial genomes. *Nucl. Acids Res.* 19, 6823–6831.

Vieu, J.F., Minck, R. and Bergogne-Bérézin, E. (1979). Bactériophages et lysotypie de Acinetobacter. *Ann. Microbiol.* 130A, 405–406.

Vila, J., Almela, M. and Jimenez de Anta, M.T. (1989). Laboratory investigation of hospital outbreak caused by two different multiresistant *Acinetobacter calcoaceticus* subsp. *anitratus* strains. *J. Clin. Microbiol.* 27, 1086–1089.

Wagner, M., Erhardt, R., Manz, W., Amann, R., Lemmer, H., Wedi, D. and Schleifer, K.H. (1994). Development of an rRNA-targeted oligonucleotide probe specific for the genus *Acinetobacter* and its application for in situ monitoring in activated sludge. *Appl. Environ. Microbiol.* 60, 792–800.

Wayne, L.G., Brenner, D.J., Colwell, R.R., Grimont, P.A.D., Kandler, O., Krichevsky, M.I., Moore, L.H., Moore, W.E.C., Murray, R.G.E., Stackebrandt, E., Starr, M.P. and Trüper, H.G. (1987). Report of the *ad hoc* committee on reconciliation of approaches to bacterial systematics. *Int. J. Syst. Bacteriol.* 37, 463–464.

Weaver R.E. and Actis L.A. (1994). Identification of *Acinetobacter* species. *J. Clin. Microbiol.* 32, 1833.

Weernink, A., Severin, W.P.J., Tjernberg, I. and Dijkshoorn, L. (1995). Pillows, an unexpected source of *Acinetobacter*. *J. Hosp. Infect.* 29, 189–199.

Zabeau, M. and Vos, P. (1992). Selective Restriction Fragment Amplification: A General Method for DNA Fingerprinting. European Patent Office, Publication 0 534 858 A1.

Chapter 4

Epidemiology of *Acinetobacter* spp.: Surveillance and Management of Outbreaks

M.-L. Joly-Guillou and C. Brun-Buisson

CONTENTS

0-8493-9223-3/95/$0.00+$.50

4.1 INTRODUCTION

Acinetobacter spp. are ubiquitous organisms in nature, and can be obtained easily from soil, water and a variety of foods, including dairy products, poultry and frozen foods. *Acinetobacter* spp. have also been known for many years as opportunistic pathogens, but have been familiar to only a few enthusiastic microbiologists because of their relatively low virulence and confused taxonomy. However, during the last 20 years, a large number of reports of hospital-acquired infections involving *Acinetobacter* spp. have drawn attention to the increasing problem caused by these organisms.

During the early 1960s, hospital infections caused by *Mima polymorpha* and *Herellea vaginicola* were described in a series of publications (Graber et al., 1962; Robinson et al., 1964; Green et al., 1965). Previously, organisms named *Bacterium anitratum* or "B5W" had been reported sporadically in the literature as causative agents of primary human infections (Rohea and Guze, 1957; Olafsson et al., 1958). Taplin et al. (1963) reported one of the first epidemiological studies of the *Mima* and *Herellea* groups, and found that these organisms formed part of the normal skin flora of 25% of normal males. Other authors have reported that up to 7% of healthy subjects may carry these organisms in their throat (Rosenthal and Tager, 1975).

Since these early studies, the contribution of *Acinetobacter* spp. to nosocomial infection has been recognised in many reports, and these organisms are now considered to be important nosocomial pathogens, particularly in intensive care units (ICUs). Any epidemiological study must take into account the overall level of infection with which a hospital is confronted. In Europe, *Acinetobacter* spp. account for about 9% of nosocomial infection in ICUs (Joly-Guillou et al., 1993), with a more predominant role in nosocomial pneumonia, particularly ventilator-associated pneumonia. It has been suggested that the digestive tracts of hospitalised patients are frequently colonised with *Acinetobacter* spp. (Timsit et al., 1993).

Since the taxonomy of the genus *Acinetobacter* was modified by Bouvet and Grimont (1986), *Acinetobacter baumannii* has been identified as the most prevalent species of the genus to be found in clinical specimens (Bouvet and Grimont, 1987; Traub, 1989; Joly-Guillou et al., 1991). Hospital isolates of *Acinetobacter* are resistant to an increasing number of antibiotics (Garcia et al., 1983; Joly-Guillou and Bergogne-Bérézin, 1985; see also Chapter 7), and a difficult-to-treat infection is now one of the most important problems for the clinician confronted with *Acinetobacter* nosocomial infection. Hospital outbreaks of *Acinetobacter* infection have not always been defined in terms of a single strain. The aim of this chapter is therefore to provide an integrated approach to understanding the hospital epidemiology of *Acinetobacter* spp. by defining the best research methods for identifying routes of cross-infection involved in such outbreaks. Only by understanding the local epidemiology of *Acinetobacter* infections will it be possible to instigate appropriate preventive and curative measures.

4.2 OCCURRENCE AND HABITATS OF *Acinetobacter* spp.

4.2.1 Clinical and Epidemiological Features of the New Taxonomic Scheme

As described in Chapter 2, *Acinetobacter* spp. are ubiquitous organisms that are distributed widely in nature. It has been estimated that acinetobacters may constitute 0.001% of the total heterotrophic aerobic population of soil and water (Baumann, 1968). Since they are non-fastidious organisms, they can utilise a wide variety of substrates as sole carbon and energy sources, and this characteristic expands the range of their habitats. Members of the genus can be isolated from virtually 100% of soil and freshwater samples if appropriate enrichment techniques are used (von Gravenitz, 1985), as well as from raw sewage effluent, drinking water supplies (Bifulco et al., 1989), and a variety of food and dairy products (see Chapter 2). The new taxonomic scheme devised by workers at the Pasteur Institute in Paris (Bouvet and Grimont, 1986) has resulted in the recognition of the concept that different species of *Acinetobacter* may have different habitats (Bouvet and Grimont, 1987; Tjernberg and Ursing, 1989; see also Chapter 2).

TABLE 4.1

Ecological Features of the Major *Acinetobacter* Genomic Species

Species designation	DNA group	Major ecological and clinical features
A. calcoaceticus	1	Soil; human infections?
A. baumannii	2	Humans are primary reservoir; causes 90% of clinical *Acinetobacter* infections; secondary reservoirs in the environment
sp. 3	3	Soil; clinical specimens — rare in France (<1%), but one outbreak of infection described
A. haemolyticus / sp. 6	4 6	Activated sludge; clinical specimens (ca. 3%) and the hospital environment
A. junii	5	Environment; clinical specimens (<1%)
A. johnsonii / *A. lwoffii*	7 8, 9	Human skin (hands of hospital staff); clinical samples (ca. 3%); animals and animal products; soil; activated sludge
spp. 9, 10, 11	9, 10, 11	Poorly defined
A. radioresistens	12	Cotton; soil; irradiated poultry; clinical specimens; material in the hospital environment
spp. 13–17	13–17	Poorly defined

This association between different species and particular habitats is outlined below and summarised in Table 4.1.

The normal habitat of ***Acinetobacter calcoaceticus*** *sensu stricto*, as described originally by Beijerinck (1911), is soil. This species is seldom associated with human infection, and the identification of the strains responsible for the rare cases of infection described in the literature has been discussed by Tjernberg and Ursing (1989).

Acinetobacter baumannii is the species associated most commonly with nosocomial infection (84–90% of *Acinetobacter* isolates from patients; Joly-Guillou et al., 1991). No natural habitat other than man has been recognised, and the presence of this species in the hospital environment is considered to be a result of contamination from an infected or colonised patient (Bouvet and Grimont, 1987).

***Acinetobacter* sp. 3** has been found in both soil and clinical samples. This species has been associated with infection only rarely in France, but has been implicated more frequently in Sweden (Tjernberg and Ursing, 1989) where it was responsible for an outbreak involving 26 patients (Dijkshoorn et al., 1993).

Haemolytic species, such as ***Acinetobacter haemolyticus*** and ***Acinetobacter* sp. 6** have been isolated occasionally from patients, the hospital environment and activated sludge samples (Bouvet and Grimont, 1990).

Acinetobacter junii has been isolated both from patients and the environment.

Acinetobacter johnsonii and ***Acinetobacter lwoffii*** have been isolated from the skin of hospitalised patients and the hands of uninfected personnel. Several authors have suggested that the human skin is one of the natural habitats of these species. These two species have also been isolated from animals and animal products, soil and activated sludge samples, but only rarely from infected clinical specimens. *A. johnsonii* cannot grow at 37°C, and its prevalence may therefore be understated unless samples are also incubated at 30°C.

Acinetobacter radioresistens has been isolated from a range of samples, including irradiated poultry, cotton and soil (Nishimura et al., 1988). More recently, this species has also been isolated from patients (Tjernberg and Ursing, 1989; Christensen et al., 1991).

The natural habitats of other identified *Acinetobacter* species are currently poorly defined.

It is worth emphasising that uninfected hospital personnel may carry species other than *A. baumannii* on their hands, while patients in the same ward may be infected only with *A. baumannii*. Similarly, the isolation of species other than *A. baumannii* from blood or cerebrospinal fluid (CSF) can result from contamination by skin bacteria and must be confirmed by clinical signs of infection. Any attempted correlation between the presence of *Acinetobacter* spp. in the hospital environment and their role as opportunistic pathogens requires identification of isolates to the species level in order to reach correct epidemiological conclusions.

4.2.2 Human Carriage in Healthy Populations

As described previously, acinetobacters can form part of the bacterial flora of the skin, particularly in moist regions such as the axillae, groin and toe webs of normal individuals (Table 4.2). Cultures can usually be made from axillae, antecubital fossae and popliteal fossae. Over 30 years ago, Taplin et al. (1963) suggested that at least 25% of normal individuals carry glucidolytic species of *Acinetobacter* (*Herellea vaginicola*) on their skin, while 10% carry non-glucidolytic species (*Mima polymorpha*). These authors further suggested that the human skin could represent an important reservoir of infection for burn and surgical patients, and a source of contamination for blood or CSF cultures, and emphasised the importance of preoperative skin sterilisation procedures for controlling infection by these organisms. *Acinetobacter* spp. have also been found occasionally in the oral cavity and respiratory tract of healthy adults (7% of subjects with glucidolytic species and 2% with non-glucidolytic species) (Rosenthal and Tager, 1975; Glew et al.,

TABLE 4.2

Carriage of *Acinetobacter* by Normal Populations and Hospitalised Patients in Non-Outbreak Situations

Site of carriage	Normal subjects %	Normal subjects in tropics %	Hospitalised patients %
Skin			
Antecubital fossae	15[h]	ND	37[b]
			8[a]
			2[c]
Popliteal fossae	ND	ND	21[a]
Axillae	10[h]	17[h]	35[a]
	20[g]		17[c]
Toe webs	ND	20[h]	33[a]
Nose	ND	ND	57[a]
Oral cavity	7[e]	ND	57[a]
			20[c]
Trachea	ND	ND	45[h]
Rectum	0[d]	ND	0
			33[a]
			40[f]
			24[h]
			17[c]

Note: ND, not demonstrated.

References: [a]Allen and Green (1987); [b]French et al. (1980); [c]Joly-Guillou et al. (unpublished data); [d]Rosenthal (1974); [e]Rosenthal and Tager (1975); [f]Sakata et al. (1989); [g]Somerville and Noble (1970); [h]Taplin et al. (1963).

1977). The digestive tract has been reported to be a reservoir in hospitalised patients, but no digestive carriage has been described outside of hospitals.

4.2.3 Seasonal Pattern of Infections

In countries where there is a well-defined summer and winter, with a marked temperature swing, it seems that acinetobacters are isolated more commonly in the summer months (Kloos and Musselwhite, 1975). This may be a result of greater sweating in the summer, especially by males. This possibility is supported by the more frequent description of community-acquired pneumonia in developed tropical countries (Barnes et al., 1988; Anstey et al., 1992). However, no significant difference in human carriage has been observed (Table 4.2) between temperate and tropical climates. Retailliau et al. (1979) noted that *Acinetobacter anitratus* presented a seasonal pattern of infection, whereas no such difference was observed with *A. lwoffii*.

4.3 APPLICATION OF TYPING SYSTEMS IN HOSPITAL EPIDEMIOLOGY

In the last 30 years, outbreaks of *Acinetobacter* infection have been an increasing concern in hospitals. A number of phenotypic and molecular typing systems have been devised and applied to the investigation of hospital outbreaks (see Chapter 3). The use of most of these methods is limited to a few specialised centres. Only antibiogram (resistance) typing and biotyping can be used in all diagnostic laboratories. As described in Chapter 3, molecular techniques such as plasmid typing, ribotyping, pulsed-field gel electrophoresis (PFGE) and polymerase chain reaction (PCR) fingerprinting have great potential for tracing the dissemination of epidemic strains. However, in comparison with phenotypic systems, these molecular methods are generally more costly, labour-intensive and time-consuming. PCR fingerprinting appears to be a particularly promising method because of its simplicity, speed and adjustable discriminatory power (Struelens et al., 1993), but requires further investigation to optimise its reproducibility and discrimination. PFGE analysis with a combination of two enzymes (*Apa*I, *Sma*I or *Nhe*I) has proven to be stable, reproducible and more discriminatory than PCR fingerprinting or ribotyping, but the necessary equipment is costly. Details of the main features of the typing methods used for epidemiological investigations can be found in Chapter 3.

In practice, as outlined in Chapter 3, it does not seem to be appropriate to introduce typing, particularly genomic typing, of *Acinetobacter* spp. in all diagnostic laboratories. Typing methods are tools for establishing the source and modes of *Acinetobacter* transmission during outbreaks or endemic periods of infection. Bacterial surveillance and screening of resistance phenotypes seem to be useful for early detection of a possible outbreak, even if they do not allow outbreak and sporadic cases to be distinguished. Resistance phenotype analysis and biotyping are good epidemiological tools for suspecting possible *Acinetobacter* cross-infection, but only the combination of these methods with an appropriate powerful genotypic technique will allow the identity or non-identity of strains to be confirmed.

4.4 VIRULENCE AND PATHOGENICITY

Knowledge of factors influencing the virulence of *Acinetobacter* spp. is still at an elementary stage. *Acinetobacter* is known to be a pathogen of relatively low virulence, but has been documented as causing community and nosocomial infections, particularly in compromised hosts (see Chapter 5). This section summarises the few investigations that have focused on possible virulence and pathogenicity factors.

4.4.1 Virulence Factors

Hydrophobicity — This has been implicated in non-specific adherence of bacteria, and is determined by non-specific factors that include the physicochemical properties of the bacterial cell surface. Boujaafar et al. (1990) demonstrated that hydrophobicity was higher in *A. baumannii* strains isolated from infected catheters and tracheal devices than in strains isolated from the environment.

Polysaccharide capsule — The invasiveness of bacteria may be associated with surface components that protect bacteria from phagocytosis. It has been shown that decapsulation of *Acinetobacter* strains can decrease their adherence to hydrocarbons and epithelial cells by making their outer surface more hydrophilic (Rosenberg et al., 1983).

Fimbriae — Thin fimbriae about 3 nm in diameter and thick fimbriae about 5 nm in diameter have been observed in *Acinetobacter* spp. by Henriksen and Blom (1975). Whereas the presence of thick fimbriae has been correlated with twitching motility, thin fimbriae are associated with adherence to hydrocarbons (Rosenberg et al., 1982; Pines and Gutnick, 1984). The presence of fimbriae has been linked to pathogenesis in various strains, attesting to the importance of bacterial adherence for colonisation of host tissues.

Enzymic activities — A wide range of extracellular enzymes are produced by *Acinetobacter* spp. (see Chapter 2), but only a few are related to the virulence of strains. Enzymic activities of *Acinetobacter* spp. have been studied by Poh and Loh (1985) with the rapid API-ZYM system. A significant production of certain enzymes, including butyrate esterase (C-4), caprylate esterase (C-4) and leucine arylamidase, was reported in all isolates tested. These enzymes are involved in the hydrolysis of short chain fatty acids, and could therefore be responsible for damage to tissue lipids. Picard and Goullet (1990) and Pouëdras et al. (1992) have used the capacity of *Acinetobacter* to produce esterases as a typing method for this genus.

Iron acquisition — The ability of a bacterium to obtain the necessary iron for growth in the human body is an important virulence determinant. Some strains of *Acinetobacter* have been shown to produce siderophores, such as aerobactin, and iron-repressible outer-membrane receptor proteins (Smith et al., 1990; Echenique et al., 1992; Actis et al., 1993). However, no aerobactin-producing strains were identified among 50 *Acinetobacter* clinical isolates studied by Martinez et al. (1987).

Lipopolysaccharide (LPS) — In common with other Gram-negative bacteria, *Acinetobacter* spp. produce an LPS responsible for lethal toxicity in mice, pyrogenicity in rabbits, and a positive reaction in the limulus amoebocyte lysate test. The production of endotoxin *in vivo* is probably responsible for the disease symptoms observed during *Acinetobacter* septicaemia.

Slime production — Virulence is enhanced in strains of *Acinetobacter* that produce slime (Obana et al., 1985). The chemical composition of the slime produced by *Acinetobacter* is as follows: a high sugar content, a DNA content of about 3%, and an unknown substance that gives a positive orcinol reaction, but only about 70% of the total composition has so far been determined. It differs from the slime produced by *Pseudomonas aeruginosa*, of which the main component is DNA. Only a small proportion of acinetobacters are slime-producing; indeed, of the 100 isolates tested by Obana (1986), only 14 had slime-producing ability. The same author demonstrated that slime was associated with cytotoxicity against neutrophils and inhibition of the cellular migration of neutrophils into peritoneal exudate of mice. However, no correlation was observed between the amount of slime produced and the degree of virulence.

4.4.2 Pathogenicity: Experimental Studies in Mice

The virulence of clinical isolates of *Acinetobacter* spp. has been studied in a mouse model of acute systemic infection induced by intraperitoneal injection (Obana et al., 1985; Obana, 1986). An inoculum of at least 10^6 cfu was required to kill a mouse, but LD_{50} values were considerably lower (10^3–10^4 cfu/mouse) in the presence of hog gastric mucin. The virulence of some strains in this model was found to be equivalent to that of *Escherichia coli*, *Serratia marcescens* and *P. aeruginosa*, and to be greater than that of other glucose non-fermentative Gram-negative bacteria such as *P. cepacia* or *X. maltophilia*. It was shown experimentally that mixed infections of *Acinetobacter* with other bacteria resulted in greater virulence than infection with a pure culture, and that the slime of *Acinetobacter* enhanced the virulence of *E. coli*, *S. marcescens* and *P. aeruginosa*.

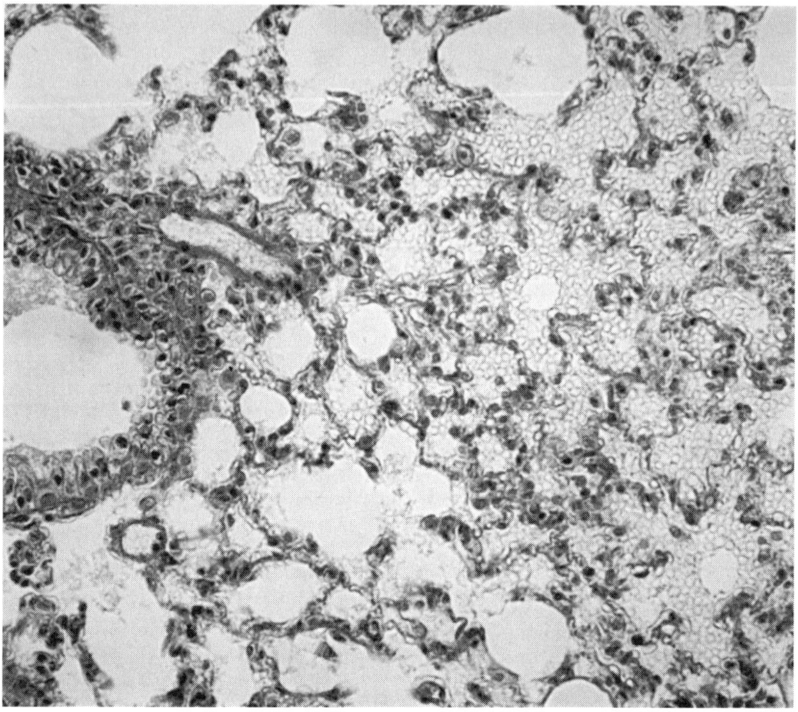

FIGURE 4.1
Mouse model of *Acinetobacter* pneumonia (day 1) showing haemorrhagic and oedematous alveolitis.

A mouse model of *Acinetobacter* pneumonia has been developed by M.-L. Joly-Guillou and M. Wolff (unpublished data). Neutropenic animals were inoculated intratracheally with 10^7 cfu/mouse. The duration of the resulting disease was 3–4 days, i.e., similar to that described for community-acquired *Acinetobacter* pneumonia in humans (Anstey et al., 1992), and the cumulative mortality was 70%. Lung pathological findings were characterised by pneumonitis with oedema and patchy distribution of haemorrhagia on day 1 (Figure 4.1), followed by purulent bronchopneumonia on day 2 (Figure 4.2). Abscess formation occurred on day 3–4. After 3–4 days, subacute pneumonitis characterised by alveolar macrophage proliferation and areas of fibrosis was observed (Figure 4.3). The bacterial count reached a mean level of 10^9 cfu on day 1–2. Since nowadays the pulmonary tract is the most frequent site of *Acinetobacter* nosocomial infection, especially in intensive care patients with ventilatory equipment, it is hoped that this model will

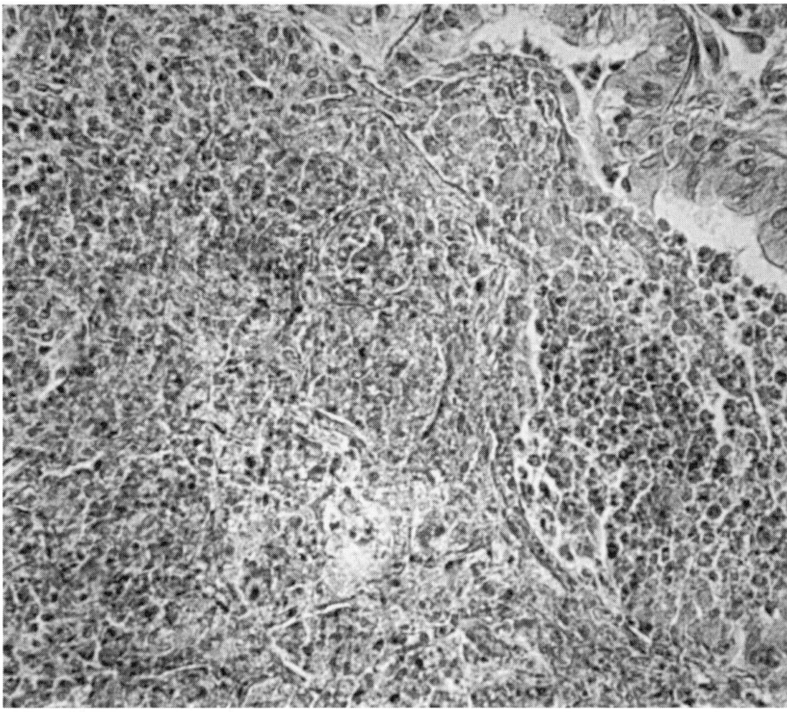

FIGURE 4.2
Mouse model of *Acinetobacter* pneumonia (day 2) showing purulent bronchopneumonia with bronchial lumen containing numerous leukocytes.

facilitate studies on the host–pathogen interactions that influence the emergence of *Acinetobacter* pulmonary infection, as well as the virulence factors involved in the pathogenesis of *Acinetobacter* disease.

4.5 ROLE OF *Acinetobacter* spp. IN HUMAN DISEASE

Acinetobacter spp. are opportunistic pathogens of relatively low virulence that cause infection only rarely in otherwise healthy individuals (see Chapter 5). Occasional sporadic cases have been described in non-immunosuppressed patients exposed to environmental sources. Most cases of infection are of nosocomial origin, and occur in patients with various conditions that predispose to colonisation and infection by strains originating from a hospital reservoir, often during epidemic outbreaks which can be linked to an inanimate reservoir.

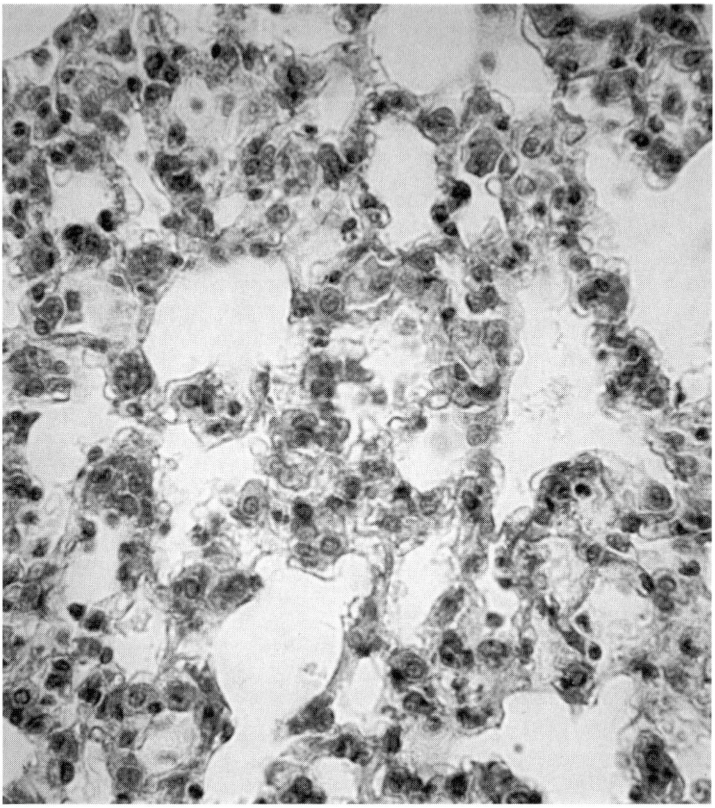

FIGURE 4.3
Mouse model of *Acinetobacter* pneumonia (day 3–4) showing alveolar macrophage proliferation.

4.5.1 Community-Acquired Infections

In a 5-year retrospective review of *Acinetobacter* bacteraemia in a 450-bed tertiary care hospital, Smego (1985) found 27 patients with positive blood cultures, of which only 18 were considered to be clinically significant. Of these 18 significant cases, only two (both caused by *A. lwoffii*) were probably community-acquired infections, but one of these episodes of bacteraemia, secondary to a urinary tract infection, occurred in a patient with bladder carcinoma. Similarly, Ramphal and Kluge (1979) considered that only 2 of 15 patients with *Acinetobacter* bacteraemia had acquired their infection in the community. Therefore, it seems that 5–10% of bacteraemias caused by *Acinetobacter* spp. may be community acquired. It is worth noting that approximately one third of all isolates recovered from blood cultures of hospitalised patients in this study were judged to be contaminants.

In a survey of healthy subjects attending various clinical screening programmes in Massachusetts, Rosenthal (1974) found that up to 6% of skin, throat or rectal swabs were positive for *Acinetobacter* carriage, with the highest carriage rates found amongst patients in nursing homes. Skin carriage rates of 20–25% have been reported (Taplin et al., 1963; Somerville and Noble, 1970). Thus, relatively low carriage rates are found amongst healthy individuals, with the skin being the predominant site of colonisation. Community-acquired infections in otherwise healthy subjects are distinctly rare. Most reports consist of single cases involving the respiratory tract (Glick et al., 1959; Reynolds and Cluff, 1963; Wands et al., 1973; Goodhart et al., 1977; Rudin et al., 1979) or the urinary tract (Hoffman et al., 1982). An interesting report describes a community outbreak of *Acinetobacter* pneumonia in foundry workers (many suffering from silicosis) who inhaled iron and silica particles that were possibly contaminated with a single strain of *Acinetobacter* (Cordes et al., 1981). The organism was isolated subsequently from air and other environmental samples in the foundry, suggesting that airborne transmission of the organism was occurring. It is likely that the underlying pulmonary condition of these patients favoured the occurrence of pneumonia.

4.5.2 Hospital-Acquired Infections

Acinetobacter spp. have received increasing recognition as nosocomial pathogens over the past 15 years. According to the recent EPIIC study (a 1-day prevalence survey of nosocomial infections in ICUs in Europe), these organisms now account for as much as 10% of the nosocomial bacteraemias recorded in ICUs. Many of the outbreaks described in recent years have been linked to environmental sources in the hospitals involved. The upsurge in the importance of *Acinetobacter* spp. as nosocomial pathogens may be linked to several different factors, including: (1) the widespread resistance to antimicrobial agents that is found in these organisms (see Chapter 7), which favours their selection in the hospital environment or a patient's own bacterial flora under the pressure of an ever-increasing consumption of antibiotics by hospitalised patients; (2) the ability of these organisms to colonise skin and mucous membranes of humans; and (3) the ability of these organisms to survive in the hospital environment for long periods, even under unfavourable conditions (see Section 4.6.1).

Prevalence and Role of Acinetobacter *spp. in Hospital Infections*

Acinetobacter spp. currently represent 2–5% of all isolates recovered in hospital clinical microbiology laboratories. Infections caused by *Acinetobacter* spp. are caused predominantly by *A. baumannii* (approximating to the former *A. anitratus* ssp. *anitratus*), while infections caused by other *Acinetobacter* spp. are relatively infrequent, with the exception of some rare outbreaks. Reports from France indicate that the proportion of *A. baumannii* isolates has increased from 75% of all *Acinetobacter* hospital isolates in the decade 1970–1980 to 90–95% of *Acinetobacter* isolates in the decade 1980–1990 (Joly-Guillou and Bergogne-Bérézin, 1985; 1990). These changes were associated with an increase in the proportion of isolates that were resistant to ß-lactams and aminoglycosides. In 1990, 84% of isolates were multiresistant to ß-lactams and 64% of isolates were resistant to aminoglycosides, although it should be noted that >50% of isolates in this survey were of two distinct phage types, indicating a high proportion of epidemic strains. In hospitals participating in the NNIS surveillance programme in the U.S., acinetobacter accounted for 0.76% of nosocomial respiratory tract infections in ICUs in the years 1974–1977, and for 6.4% in the years 1985–1990 (Jarvis and Martone, 1992). Struelens et al. (1993) reported that *A. baumannii* accounted for 0.3% of endemic nosocomial infection in ICU patients, and for 1% of nosocomial bacteraemias. An increase in nosocomial infections caused by *Acinetobacter* spp. has been recorded during the summer months in several reports (Ramphal and Kluge, 1979; Retailliau et al., 1979). This seasonal increase has not been explained satisfactorily, although temperature and humidity conditions during the summer are possible predisposing factors for overgrowth of *Acinetobacter* in the environment or in patients.

Colonisation and Infection

A large proportion of clinical isolates are associated with colonisation rather than infection of patients. Studies of carriage in the hospital population have shown that *Acinetobacter* spp. may be found as normal inhabitants on skin and in the oropharynx and digestive tract of hospitalised patients in a much higher proportion than in non-hospitalised patients. Rosenthal (1974) found that throat swabs were positive for *Acinetobacter* spp. in 7–18% of patients, while tracheostomy swabs were positive in 45% of hospitalised patients. High colonisation rates of the skin, throat, respiratory system or digestive tract, of varying degrees of importance, have been documented in several outbreaks. A number of conclusions regarding colonisation in hospitalised patients can be drawn from these studies: (1) a high rate of colonisation can be found in debilitated hospitalised patients, especially during outbreak situations; (2) a predominant site of colonisation is the skin, but other sites, such as the respiratory or digestive tracts, may also be involved and may predominate on certain occasions; and (3) the discrepancy between carriage rates for outpatients and hospitalised patients suggests that infecting or colonising organisms in hospital-acquired infections may derive more often from cross-transmission or environmental sources rather than from endogenous sources in patients.

The large proportion of colonised patients in a given hospital setting means that the differentiation between colonisation and infection may not be straightforward. Nosocomial infections may involve any site, but predominate in the respiratory tract, urinary tract and wounds. The trend observed in recent years is towards an increasing clinical significance of *Acinetobacter* isolates recovered from clinical specimens. Data from the NNIS indicate that 45% of infections involving *Acinetobacter* spp. are actually polymicrobial. The overall rate of secondary bacteraemia was 4.3%, but infections at sites other than those mentioned above were the sites associated with the highest rates of secondary bacteraemia (Retailliau et al., 1979), with infections of burned skin, the central nervous system and intra-abdominal infections being associated with bacteraemia in 20–25% of cases, whereas isolates from the more classical sites of nosocomial infection were associated with bacteraemia in only 0.5–5% of cases. However, more recent data from the CDC have linked 16.2% of *Acinetobacter* nosocomial isolates with secondary bacteraemia (Horan et al., 1986). Many isolates from skin and the respiratory tract should still be considered as colonising rather than infecting organisms. At Bichat Hospital in Paris, a steady increase in the proportion of *Acinetobacter* isolates from superficial wounds has been recorded over the past decade (increasing from 25 to 45%), reflecting the growing role of skin colonisation (Joly-Guillou and Bergogne-Bérézin, 1990), whereas isolates recovered from urine samples decreased proportionately from 35 to 16%, and 11.5% of isolates were recovered from blood cultures. While many isolates from the respiratory tract and superficial wounds probably again represent colonising organisms, these two sources, and especially the skin, serve as important reservoirs of infecting organisms during outbreak situations.

Risk Factors for Hospital-Acquired Infection

A large proportion of clinical *Acinetobacter* isolates originate from patients hospitalised in surgical wards and ICUs. Joly-Guillou and Bergogne-Bérézin (1990) found that 57% of 1056 isolates were recovered from patients in ICUs. In the NNIS study (Retailliau et al., 1977), 62% of isolates originated from patients in surgical wards. Buxton et al.

(1973) found that most infected or colonised patients were high-risk surgical patients, burn patients or ICU patients. In common with other antibiotic-resistant Gram-negative bacilli, an extended hospital stay, instrumentation at various sites, surgical procedures and previous receipt of antibiotics are common risk factors for infection with *Acinetobacter* spp. (Ramphal and Kluge, 1979). In a retrospective case-control study, Peacock et al. (1988) identified three risk factors for respiratory tract colonisation or infection with aminoglycoside-resistant acinetobacters; i.e., extended ICU care, prolonged respiratory therapy, and previous receipt of cephalosporins and aminoglycosides. Struelens et al. (1993) identified emergency surgery, mechanical ventilation and previous treatment with broad-spectrum penicillins as risk factors during a small outbreak.

4.6 HOSPITAL EPIDEMIOLOGY OF *Acinetobacter* spp.

4.6.1 Reservoirs of *Acinetobacter* spp. in the Hospital Environment

Numerous studies have documented the presence of *Acinetobacter* spp. in the hospital environment. However, rates of positive cultures may vary widely, depending on the epidemiological setting. Rosenthal (1974) found *Acinetobacter* spp. in 27% of hospital sink traps and 20% of hospital floor swab cultures. Air contamination in the absence of a colonised patient is comparatively rare. Gram-negative bacteria do not normally survive for long in air following drying (Ayliffe and Lowbury, 1982), and are not spread by the aerial route except when dispersed in fine liquid droplets, such as are produced in aerosols (Smith and Massanari, 1977; Craven et al., 1984). However, several studies have documented extensive contamination by *Acinetobacter* spp. of the environment, including air samples, in the vicinity of infected or colonised patients. For example, during an outbreak of infection originating from an ICU, 12 (11.5%) of 104 air samples from wards harbouring colonised patients were positive for *Acinetobacter* spp., as well as 7 (8%) of 89 sink-trap swabs and 13 (17%) of 75 samples from bedside cupboards in the same areas (Crombach et al., 1989); however, <2% of samples taken from areas without colonised or infected patients were positive. Likewise, in an outbreak of *Acinetobacter* septicaemia in a neonatal ICU, *Acinetobacter* spp. were found in 15% of swab samples from sinks used for handwashing (Sakata et al., 1989), and it should be noted that the strains persisted at these sites even after control of the outbreak.

Extensive contamination of the environment, including air samples, was found in an outbreak of infection with multiresistant *Acinetobacter* spp. (Allen and Green, 1987); again, contamination was found essentially in the vicinity of infected or colonised patients, with 16 of 82 settle-plate cultures from the environment of three colonised patients found to be positive. Bed linen from colonised patients was positive consistently, but linen from non-colonised patients was also positive on several occasions, as well as over-blankets from empty beds in the vicinity of one ventilated colonised patient, and bed curtains around colonised patients. Persistent contamination was documented for up to 13 days after the discharge of a patient.

Survival in the Environment

The above data indicate that certain *Acinetobacter* spp., unlike other Gram-negative bacteria, can persist in the environment for several days, even in dry conditions, on particles and dust, thereby probably contributing to the development and persistence of outbreaks. The survival of *Acinetobacter* spp. on dry surfaces has been studied by Getchell-White et al. (1989) and Musa et al. (1990). The latter authors found that *A. calcoaceticus*, in contrast to *A. lwoffii*, could survive on dry surfaces for up to 60 h, while Getchell-White et al. (1989) reported that *A. calcoaceticus* and *A. lwoffii* could persist for means of 8.2 and 10.2 days, respectively; i.e., durations even longer than that found for *S. aureus*. Environmental contamination during an outbreak in a paediatric ICU was studied extensively and was demonstrated on various equipment and surfaces in the unit (telephone handles, door pushplates, patient's charts, table tops, etc.), all of which were probably contaminated by the hands of staff, which were also shown to be contaminated. Allen and Green (1987) found that an epidemic strain of multiresistant *Acinetobacter* spp. could survive for up to 6 days after inoculation onto dry filter paper, a duration similar to that found with *S. aureus*, which persisted for 7 days, but in contrast to *E. coli* and *Pseudomonas* spp., which survived for 24 h or less.

More recently, Wendt et al. (1994) tested the ability of *Acinetobacter* spp. (clinical isolates and ATCC strains) to survive on dry surfaces found in the hospital environment (PVC, ceramics, rubber and steel) in comparison with *E. coli* and *Enterococcus faecium*. From an inoculum of 10^8 cfu/ml, three of the five *Acinetobacter* clinical isolates, which were heavily encapsulated, survived for >4 months on nearly all of the materials tested. All other strains tested died within 8 weeks. It was suggested that the capsule could enable survival in conditions that were hostile for the organism. Prolonged survival of *Acinetobacter* spp. has also been described previously on hospital floors and air-dried washcloths (Buxton et al., 1978).

It has been reported that *Acinetobacter* spp. can survive exposure to chlorhexidine, gluconate and phenol-based disinfectants, particularly if the exact recommended concentrations are not strictly adhered to (Berry et al., 1990). The presence of contaminating organic matter was shown to favour the survival of *Acinetobacter* spp. in the presence of disinfectants. In addition, radiation resistance of *Acinetobacter* clinical isolates was demonstrated by Christensen et al. (1991). Radiation resistance has been described previously for genomic species 12 (*A. radioresistens*), but some other species, including *A. baumannii* and *A. lwoffii*, were shown in this study to have increased their radiation resistance compared with similar isolates collected in 1970. These results indicate that special attention should be paid to medical devices that are normally sterilised by irradiation, and particularly devices used in ICUs.

In summary, *Acinetobacter* spp. have unique characteristics among nosocomial Gram-negative bacteria that favour their persistence in the hospital environment. These organisms spread easily in the environment of infected or colonised patients, and can persist in that environment for many days, a factor which may explain their propensity for causing extended outbreaks.

Contaminated Materials: A Source of Outbreaks

The dissemination and persistence of *Acinetobacter* spp. in the hospital environment probably accounts for the specific role that has been reported for contaminated materials as a reservoir of infection, especially during outbreaks. In some persistent outbreaks, contaminated materials have been shown to act as a *source* of the outbreak. Materials used for respiratory therapy or support have been implicated in many such cases (Smith and Massanari, 1977; Hartstein et al., 1988; Cefai et al., 1990). Thus, an outbreak of 24 cases of infection with *Acinetobacter* spp., occurring mostly in debilitated patients with intravascular catheters, was traced to contaminated room-air humidifiers (Smith and Massanari, 1977). Contaminated air samples were found up to 10 m from the humidifiers, and probable skin colonisation of patients in the vicinity of the devices resulted in intravascular catheter infection. A similar outbreak occurred in patients receiving peritoneal dialysis (Abrutyn et al., 1978), and contamination of dialysis fluid bottles, via a contaminated water bath used to warm the dialysis fluid, was demonstrated. This outbreak was controlled after revision of the decontamination procedures for the heating buckets and for starting infusion of dialysis fluids. Other outbreaks have involved inadequate sterilisation of autoclavable reusable needles used for administration of intrathecal methotrexate in leukaemic patients (Kelkar et al., 1989), defective heating of a washing machine used for decontamination of reusable ventilator tubings (Cefai et al., 1990), and inadequate decontamination of respiratory monitoring devices and resuscitation bags (Stone and Das, 1985; Hartstein et al., 1988; Vandenbroucke-Grauls et al., 1988). Ethylene oxide sterilisation of such contaminated equipment has been shown to be effective.

However, although respiratory equipment may be responsible for persistent outbreaks as a result of inadequate decontamination between use with consecutive patients, such equipment may act in some instances as only an intermediate reservoir of organisms, and not as the primary source of infection. Indeed, Buxton et al. (1978) showed that, while various items of respiratory equipment were positive for *Acinetobacter* spp., a respiratory therapist with dermatitis hand lesions was a chronic carrier of *Acinetobacter* and was contaminating the equipment during assembly and testing. Medical equipment may therefore become contaminated both by the patients themselves and by staff during handling, and the latter possibility should always be considered when outbreaks of infection occur, especially in respiratory ICUs.

A further notable example of the role of the environment in the dissemination of *Acinetobacter* spp. was described by Sherertz and Sullivan (1985), who reported a large outbreak involving 63 of 103 patients admitted to a burns unit over a 21-month period. The outbreak was traced to contamination of mattresses through breaches in their plastic covers that allowed water penetration and subsequent persistence of the organism in the wet foam of the mattresses.

4.6.2 Human Reservoirs in Hospitals

The carriage rate of *Acinetobacter* spp. in non-hospitalised patients is low, but may be much higher in hospitalised patients, especially during outbreaks of infection (see previous section). In such instances, sites of carriage by humans may be varied; thus, while the skin and wounds are the predominant reservoirs, the respiratory tract and/or the digestive tract may also serve as reservoirs of *Acinetobacter* spp., depending on the precise epidemiological setting.

Respiratory Tract Colonisation

Outbreaks involving mechanically ventilated ICU patients are often associated with a high colonisation rate of the respiratory tract (Buxton et al., 1978; Allen and Green, 1987; Gerner-Smidt, 1987; Peacock et al., 1988), a fact which may point to contamination of respiratory therapy equipment as the possible source of an outbreak. In the study by Buxton et al. (1978) of an 8-month outbreak, 20 of 24 isolates were from the respiratory tract, and all but one were recovered from ICU patients. Similarly, in a protracted outbreak involving 111 patients in an

ICU, *Acinetobacter* spp. was recovered from the respiratory tract in 109 of the 111 patients affected (Gerner-Smidt, 1987), while Peacock et al. (1988) reported an outbreak involving 98 patients over an 18-month period in which 87% of isolates were recovered from ICU patients, and 92% were recovered from respiratory tract samples. In this latter outbreak, only one of 136 environmental cultures was positive for *Acinetobacter* spp., but the organism was found on the hands of staff after sampling with the sterile bag technique. Allen and Green (1987) described a 10-month hospital outbreak, involving 36 patients from three neurosurgical wards, one medical ward and the ICU of a general hospital, in which 77% of cases involved the respiratory tract of ventilated patients. In this last outbreak, in contrast to those described above, extensive contamination of the environment surrounding patients was found, but contamination of respiratory therapy equipment was not demonstrated.

Skin Colonisation

As mentioned previously, patients often have skin colonisation during outbreaks. When studied in a systematic fashion, at least one third of patients with clinically significant infection or colonisation at other sites were also found to be colonised at skin sites (e.g., axillae, forearms, toewebs, etc.). Skin colonisation of patients plays an important role in subsequent contamination of the hands of hospital staff during trivial contacts, thereby contributing to the spread and persistence of outbreaks (Getchell-White et al., 1989).

Digestive Tract Colonisation

Colonisation of the digestive tract of patients with *Acinetobacter* spp. is normally unusual; however, several studies have documented oropharyngeal colonisation of patients with respiratory tract colonisation, and digestive tract colonisation has been reported to be a major reservoir of resistant strains in several outbreaks. Thus, during an outbreak in a neonatal ICU, *Acinetobacter* colonisation was documented in 72% of infants, and faecal carriage occurred in 40% of infants (Sakarta et al., 1989). Substantial contamination of the ICU environment (notably sinks) and equipment was also found in this outbreak. Wise and Tosolini (1990) described a protracted outbreak involving 237 patients over a 2-year period, 57% of whom came from a single long-stay specialised spinal cord injury unit. Most (133 of 156) isolates from patients in the spinal cord unit were recovered from urine cultures and had a distinctive antibiotic resistance profile, whereas isolates from other wards were recovered mostly from wounds and respiratory tract samples. Of 79 new admissions to the spinal cord injury unit, 40% acquired digestive tract colonisation with *Acinetobacter* within 2 months, and 92% of patients who were resident in the unit for at least 5 months were colonised. Substantial contamination of the environment was also found in this study. In contrast, French et al. (1980) found no rectal carriage of *Acinetobacter* spp. in an outbreak involving several units, including a urology ward with a high rate of urinary tract infection caused by an epidemic *Acinetobacter* strain.

4.6.3 Transmission of Infection

Airborne Spread

Numerous studies have shown that dust from bedding, floors, clothes and skin is an important vehicle of airborne bacteria. Airborne transmission of respiratory tract infection can occur if virulent organisms carried on particles of a certain dimension are inhaled. A possible role of airborne spread in *Acinetobacter* nosocomial infection was suggested by Allen and Green (1987). It was demonstrated by Craven et al. (1984) that ventilator circuit medication nebulisers, when seeded with bacteria, produced aerosols containing higher numbers of *Acinetobacter* spp. than *P. aeruginosa*. It has been suggested that the morphology of *Acinetobacter* cells (similar in size and shape to staphylococci) might favour airborne spread, and that the particles carrying *Acinetobacter* cells are the same size as those carrying staphylococci (Gerner-Smidt, 1987). In a study during an endemic period of infection in an ICU (Gerner-Smidt, 1987), *Acinetobacter* was found in the air, albeit at a low concentration of 3.5×10^{-2} cfu/m^3. In a separate incident, Cordes et al. (1981) described a cluster of pneumonia cases in foundry workers; *A. calcoaceticus* was isolated from air samples and airborne spread of the organism was suggested. In conclusion, it seems that air transmission can undoubtedly contribute to the spread of endemic or epidemic infection, but perhaps only when the environment is contaminated heavily.

Hand Carriage

Infected or colonised patients seem to be the primary reservoir of *A. baumannii* in hospitals, and contaminated equipment is a frequent source of infections; however, the way(s) in which the organism spreads from patient to patient is less clear. As indicated previously, staff hand carriage from contact with an infected patient is the most likely mode of transmission, which may also involve indirect hand contamination from contact with inanimate objects. Musa et al. (1990) studied the survival of *A. baumannii* and *A. lwoffii* on human skin (Figure 4.4). There is evidence to suggest that *Acinetobacter* survives better on fingertips

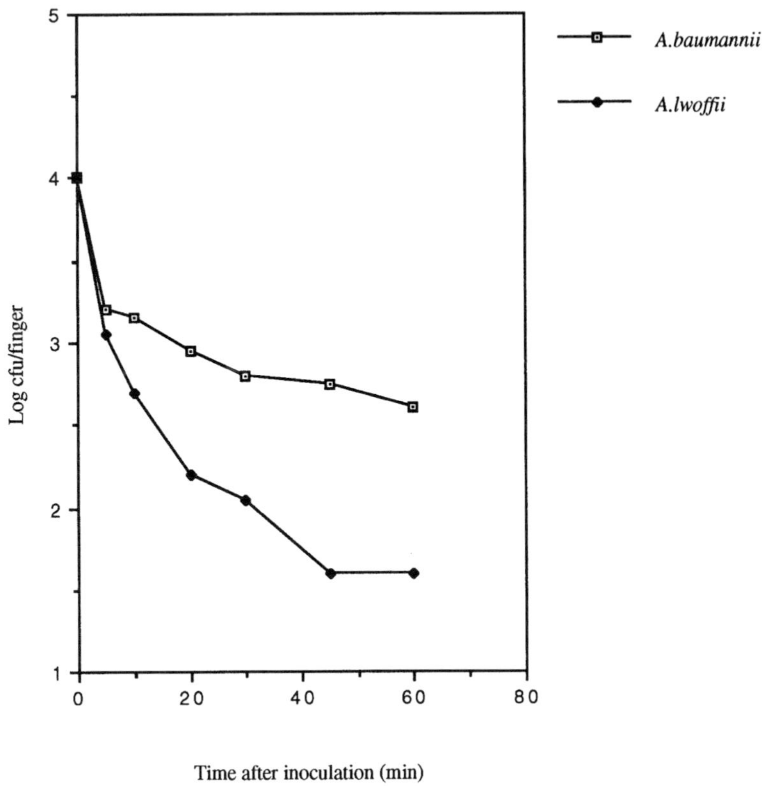

FIGURE 4.4
Survival of *A. baumannii* and *A. lwoffii* on the fingertips of volunteers (mean values calculated from the data of Musa et al., 1990).

than other Gram-negative bacilli such as *Klebsiella aerogenes, E. coli, S. marcescens* and *P. aeruginosa* (Casewell and Desai, 1983; Ayliffe et al., 1988). From an *Acinetobacter* inoculum of about 2×10^6 cfu/finger, up to 10^2 cfu/finger were recovered 1 h after inoculation. The relatively short survival time observed on fingertips, compared to survival on dry surfaces, has been attributed to antimicrobial action on the surface of palmar skin. However, even survival for 1 h on skin provides ample opportunity for transmission in the absence of appropriate hand-washing and disinfection procedures.

Hand carriage of *Acinetobacter* spp. by staff has been documented in almost all extended outbreaks investigated to date. Thus, Buxton et al. (1978) found that one third of staff had transient hand carriage, while French et al. (1980) found hand carriage by 11 of 38 staff working in wards affected by a hospital-wide outbreak. These latter authors also demonstrated persistent colonisation of hands in a health-care worker following experimental contamination. Getchell-White et al. (1989) found hand carriage by 5–50% of staff in several paediatric and neonatal ICUs having problems with *Acinetobacter* infections.

Transient hand carriage may be detected most readily by the sterile-bag culture technique (Hartstein et al., 1988; Peacock et al., 1988). The latter authors failed to detect hand carriage of aminoglycoside-resistant *Acinetobacter* spp. with swab cultures, but found a carriage rate of 28% by the sterile-bag technique; no other source in the environment was demonstrated in this outbreak. Patterson et al. (1991) reported transmission of *Acinetobacter* spp. following improper use of disposable gloves, which were not changed routinely between patient contacts. It therefore seems that contamination of hands of staff, or gloves worn by staff, may serve as an important route of transmission of *Acinetobacter* spp., the relative importance of which may vary depending on the precise epidemiological setting.

4.7 INVESTIGATION AND CONTROL OF OUTBREAKS

4.7.1 Endemic vs. Epidemic Infection

Acinetobacter spp. now account for a substantial proportion of endemic nosocomial infections. As described previously, *Acinetobacter* spp. are responsible for 0.5–1% of nosocomial infections. The relative importance of endemic and epidemic infections is difficult to delineate. Since about 50% of isolates are recovered from ICU patients, and many such isolates correspond to epidemic strains, it can be estimated that about half of all infections are of epidemic origin. The distinction between endemic and epidemic infections often requires the use of several typing techniques, some of which are not available routinely at present (see Chapter 3 and Section 4.3). As is common with epidemics of other multiresistant Gram-negative bacilli, ICUs have a predominant role in hospital outbreaks. Recent trends indicate increasing antimicrobial resistance of *Acinetobacter* isolates (Buisson et al., 1990; Joly-Guillou and Bergogne-Bérézin, 1990; Chapter 7), posing a serious threat to hospitalised patients. Such outbreaks are often found to be associated with multiple epidemic strains by sensitive typing techniques, and these techniques can help to distinguish epidemic from endemic infections, thereby orienting further epidemiological investigations; however, multiresistant strains differing by only one or two antibiotic susceptibility markers are often shown to be related closely (Go et al., 1994; Seifert et al., 1994; Tankovic et al., 1994). For practical purposes, multiresistant strains of a similar antibiotype should be considered epidemic *a priori* unless proven otherwise.

4.7.2 Approach to Management and Control of Outbreaks

The variety of potential sources of contamination or infection with *Acinetobacter* spp. in the hospital environment makes control of outbreaks caused by these organisms one of the most difficult challenges in infection control. As emphasised above, outbreaks may result from intrinsic contamination of medical equipment and devices (e.g., respiratory equipment, intravenous catheters or needles) used in patients for monitoring or therapy, and/or from contamination of the environment, either by the airborne route or by contact with patients (e.g., mattresses, air humidifiers). Persistence of *Acinetobacter* spp. in the environment provides ample opportunities for contamination of patients and staff, and may explain continuing long-term outbreaks. The emphasis of initial control measures should, however, be on strict isolation of infected or colonised patients to limit dissemination of outbreak strains in the environment. Isolation and cohorting of patients and staff has been shown to be effective in some outbreaks (French et al., 1980; Patterson et al., 1991; Go et al., 1994; Seifert et al., 1994), but this method alone has often been found to be insufficient to control outbreaks. It is useful to review hand-washing policy and practices, as cross-transmission via the hands of staff has been demonstrated in several outbreaks (Getchell-White et al., 1989; Go et al., 1994). Concomitant to isolation precautions, an investigation of the dissemination of *Acinetobacter* strains in the environment should be undertaken, and if substantial contamination is found in the vicinity of infected or colonised patients, housekeeping practices should be reviewed, and reinforced if necessary (Crombach et al., 1989). In some instances, extensive measures, including closing the unit for complete disinfection has been necessary (Sherertz and Sullivan, 1985; Tankovic et al., 1994). In one outbreak, even this measure was not sufficient because of persistence of the epidemic organism on contaminated equipment in the unit (Sherertz and Sullivan, 1985). Case-control studies are appropriate to find a common environmental source to which patients have been exposed, which may be suggested by a characteristic epidemic curve. When exposure to a particular item of equipment is implicated, it is necessary to determine whether extrinsic contamination has occurred during use in infected or colonised patients, or whether intrinsic contamination (e.g., via ineffective sterilisation or contamination by staff carriers during handling) is the cause (Hartstein et al., 1988).

There is some evidence that increased consumption of antibiotics favours the emergence and spread of *Acinetobacter* spp. (Buisson et al., 1990; Joly-Guillou and Bergogne-Bérézin, 1990). Go et al. (1994) described an outbreak of imipenem-resistant *Acinetobacter* infections following increased use of imipenem in response to an outbreak involving cephalosporin-resistant *K. pneumoniae.* Multiresistant *Acinetobacter* spp. are likely to be selected in the hospital environment in response to increasing antibiotic pressure. Control of antibiotic usage is therefore also an important part of preventive measures against the emergence of epidemic *Acinetobacter* infection.

Acinetobacter spp. are increasingly important nosocomial pathogens, and are capable of rapid adaptation to the hospital environment. There is no doubt that these organisms will pose continuing problems in the future, which is disturbing because of the extent of their ever-increasing antibiotic resistance profiles. A combination of control measures is often required to contain these organisms. Continued awareness of the need to maintain good housekeeping and control of the environment, including equipment decontamination, strict attention to hand washing and isolation procedures, and control of antibiotic usage, especially in high-risk areas, appears to be the combination of measures most likely to control the previously unabated spread of *Acinetobacter* spp. in hospitals.

REFERENCES

Abrutyn, E., Goodhart, G.L., Ross, K., Anderson, R. and Buxton, A. (1978). *Acinetobacter calcoaceticus* outbreak associated with peritoneal dialysis. *Am. J. Epidemiol.* 10, 328–335.

Actis, L.A., Tolmasky, M.E., Crosa, L.M. and Crosa, J.H. (1993). Effect of iron-limiting conditions on growth of clinical isolates of *Acinetobacter baumannii. J. Clin. Microbiol.* 31, 2812–2815.

Allen, K.D., and Green, H.T. (1987). Hospital outbreak of multi-resistant *Acinetobacter anitratus*: an airborne mode of spread. *J. Hosp. Infect.* 9, 110–119.

Anstey, N.M., Currie, B.J. and Withnall, K.M. (1992). Community-acquired pneumonia in the northern territory of Australia. *Clin. Infect. Dis.* 14, 83–91.

Ayliffe, G.A.J. and Lowbury, E.J.L. (1982). Airborne infection in hospital. *J. Hosp. Infect.* 3, 217–240.

Ayliffe, G.A.J., Babb, J.R., Davies, J.G. and Lilly, H.A. (1988). Hand disinfection: a comparison of various agents in laboratory and ward studies. *J. Hosp. Infect.* 11, 226–243.

Barnes, D.J., Naraqui, S. and Igo, J.D. (1988). Community-acquired *Acinetobacter* pneumoniae in adults in Papua New Guinea. *Rev. Infect. Dis.* 10, 636–639.

Baumann, P. (1968). Isolation of *Acinetobacter* from soil and water. *J. Bacteriol.* 96, 39–42.

Beijerinck, M.W. (1911). Über Pigmentbildung bei Essigbakterien. *Zentralbl. Bakteriol. Parasit.* 29, 169–176.

Berry, R., Towner, K.J. and Slack R.C.B. (1990). Resistance of *Acinetobacter* to drying and disinfectants. In *Abstr. 2nd Int. Workshop Biology of* Acinetobacter *(Paris).*

Bifulco, J.M., Shirey, J.J. and Bissonnette, G.K. (1989). Detection of *Acinetobacter* spp. in rural drinking water supplies. *Appl. Environ. Microbiol.* 55, 2214–2219.

Boujaafar, N., Freney, J., Bouvet, P.J.M. and Jeddi, M. (1990). Cell surface hydrophobicity of 88 clinical strains of *Acinetobacter baumannii. Res. Microbiol.* 141, 477–482.

Bouvet, P.J.M. and Grimont, P.A.D. (1986). Taxonomy of the genus *Acinetobacter* with the recognition of *A. baumannii*, sp. nov., *A. haemolyticus*, sp. nov., *A. johnsonii*, sp. nov., and *A. junii*, sp. nov. and emended descriptions of *A. calcoaceticus* and *A. lwoffii. Int. J. Syst. Bacteriol.* 36, 228–240.

Bouvet, P.J.M. and Grimont, P.A.D. (1987). Identification and biotyping of clinical isolates of *Acinetobacter. Ann. Inst. Pasteur Microbiol.* 140, 291–299.

Bouvet, P.J.M., Jeanjean, S., Vieu, J.F. and Dijkshoorn, L. (1990). Species, biotype, and bacteriophage type determinations compared with cell envelope protein profiles for typing *Acinetobacter* strains. *J. Clin. Microbiol.* 28, 170–176.

Buisson Y., Tran Van Nhieu, G., Ginot, L., Bouvet, P., Schill, H., Driot, L. and Meyran, M. (1990). Nosocomial outbreaks of amikacin-resistant tobramycin-sensitive *Acinetobacter* species: correlation with amikacin usage. *J. Hosp. Infect.* 15, 83–93.

Buxton, A.E., Anderson, R.L., Werdegar, D. and Atlas, E. (1978). Nosocomial respiratory tract infection and colonization with *Acinetobacter calcoaceticus. Am. J. Med.* 65, 505–513.

Casewell, M.W. and Desai, N. (1983). Survival of multiply-resistant *Klebsiella aerogenes* and other Gram-negative bacilli on fingertips. *J. Hosp. Infect.* 4, 350–360.

Cefai, C., Richards, J., Gould, F.K. and McPeake, P. (1990). An outbreak of *Acinetobacter* respiratory tract infection resulting from incomplete disinfection of ventilatory equipment. *J. Hosp. Infect.* 15, 177–182.

Christensen, E.A., Gerner-Smidt, P. and Kristensen, H. (1991). Radiation resistance of clinical *Acinetobacter* spp. A need for concern? *J. Hosp. Infect.* 18, 85–92.

Cordes, L.G., Brink, E.W., Checko, P.J., Lentnek, A., Lyons, R.W., Hayes, P.S., Wu, T.C., Tharr, D.G. and Fraser, D.W. (1981). A cluster of *Acinetobacter* pneumonia in foundry workers. *Ann. Intern. Med.* 95, 688–695.

Craven, D., Litchenberg, D.A., Goularte, T.A., Make, B.J. and McCabe, W.R. (1984). Contaminated medication nebulizers in mechanical ventilation circuits. *Am. J. Med.* 77, 834–838.

Crombach, W.H.J., Dijkshoorn, L., van Noort-Klaassen, M., Niessen, J. and van Knippenbert-Gordebeke, G. (1989). Control of an epidemic spread of multi-resistant *Acinetobacter calcoaceticus* in a hospital. *Intensive Care Med.* 15, 166–170.

Dijkshoorn, L., Aucken, H.M., Gerner-Smidt, P., Kaufmann, M.E., Ursing, J. and Pitt, T.L. (1993). Correlation of typing methods for *Acinetobacter* isolates from hospital outbreaks. *J. Clin. Microbiol.* 31, 702–705.

Echenique, J.R., Arienti, H., Tolmasky, M.E., Read, R.R., Staneloni, R.J., Crosa, J.H. and Actis, L.A. (1992). Characterization of a high-affinity iron transport system in *Acinetobacter baumannii. J. Bacteriol.* 174, 7670–7679.

French, G.I., Casewell, M.W., Roncoroni, A.J., Knight, S. and Phillips, I. (1980). A hospital outbreak of antibiotic-resistant *Acinetobacter anitratus*: epidemiology and control. *J. Hosp. Infect.* 1, 125–131.

Garcia, I., Fainstein, V., LeBlanc, B. and Bodey, G.P. (1983). *In vitro* activities of new beta-lactam antibiotics against *Acinetobacter* spp. *Antimicrob. Agents Chemother.* 24, 297–299.

Gerner-Smidt, P. (1987). Endemic occurrence of *Acinetobacter calcoaceticus* biovar *anitratus* in an intensive care unit. *J. Hosp. Infect.* 10, 265–272.

Getchell-White, S.I., Donowitz, L.G. and Groschel, D.H.M. (1989). The inanimate environment of an intensive care unit as a potential source of nosocomial bacteria: evidence for long survival of *Acinetobacter calcoaceticus*. *Infect. Control Hosp. Epidemiol.* 10, 402–406.

Glew, R.H., Moellering, R.C. and Kunz, L.J. (1977). Infections with *Acinetobacter calcoaceticus* (*Herellea vaginicola*): clinical and laboratory studies. *Medicine* 56, 79–97.

Glick, L.M., Moran, G.P., Coleman, J.M. and O'Brien, G.F. (1959). Lobar pneumonia with bacteremia caused by *Bacterium anitratum*. *Am. J. Med.* 27, 183–186.

Go, E.S., Urban, C., Burns, J., Kreiswirth, B., Eisner, W., Mariano, N., Mosinka-Snipas, K. and Rahal, J.J. (1994). Clinical and molecular epidemiology of *Acinetobacter* infections sensitive only to polymyxin B and sulbactam. *Lancet* 344, 1329–1332.

Goodhart, G.L., Abrutyn, E., Watson, R., Root, R.K. and Egert, J. (1977). Community-acquired *Acinetobacter calcaceticus* var. *anitratus* pneumonia. *J. Am. Med. Assoc.* 238, 1516–1518.

Graber, C.D., Rabin, E.R., Mason, E.D. and Vogel, E. (1962). Increasing incidence of nosocomial *Herellea vaginicola* infections in burned patients. *Surg. Gynecol. Obstet.* 114, 109–112.

Green, G.S., Johnson, R.H. and Shively, J.A. (1965). *Mimeae*: opportunistic pathogens. *J. Am. Med. Assoc.* 194, 163–166.

Hartstein, A.I., Rashad, A.L., Lieber, J.M., Actis, L.A., Feeman, J., Rourke, J.W., Stibolt, T.B., Tomalsky, M.E., Ellis, G.R. and Crosa, J.H. (1988). Multiple intensive care outbreak of *Acinetobacter calcoaceticus* subspecies *anitratus* respiratory infection and colonization associated with contaminated reusable ventilator circuits and resuscitation bags. *Am. J. Med.* 85, 624–631.

Henriksen, G.I. and Blom, J. (1975). Correlation between twitching motility and possession of polar fimbriae in *Acinetobacter calcoaceticus*. *Acta Pathol. Microbiol. Scand.* B83, 103–108.

Hoffmann, S., Mabeck, C.E. and Vejsgaard R. (1982). Bacteriuria caused by *Acinetobacter* biovars in a normal population in general practice. *J. Clin. Microbiol.* 16, 443–451.

Horan, T.C., White, J.W. and Jarvis, W.R. (1986). Nosocomial infection surveillance, 1984. *Morbid. Mortal. Weekly Rep.* 35, 17.

Jarvis, W.R. and Martone, W.J. (1992). Predominant pathogens in hospital-acquired infection. *J. Antimicrob. Chemother.* 29 (Suppl A), 19–27.

Joly-Guillou, M.-L. and Bergogne-Berezin, E. (1985). Evolution d'*Acinetobacter calcoaceticus* en milieu hospitalier de 1971 à 1984. *Presse Med.* 14, 2331–2335.

Joly-Guillou, M.-L. and Bergogne-Berezin, E. (1990). Epidémiologie et résistance aux antibiotiques d'*Acinetobacter* en milieu hospitalier: bilan de 5 années. *Presse Med.* 19, 357–361.

Joly-Guillou, M.-L., Bergogne-Berezin, E. and Vieu, J.F. (1991). Epidemiology of *Acinetobacter* strains isolated from nosocomial infections in France. In *The Biology of* Acinetobacter, pp. 63–68. K.J. Towner, E. Bergogne-Bérézin and C.A. Fewson (Eds). Plenum Press, New York.

Joly-Guillou, M.-L., Decré, D., Wolff, M. and Bergogne-Berezin, E. (1993). Infections nosocomiales à *Acinetobacter*, bilan d'un enquête retrospective dans les unités de soins intensifs des CHU français en 1991. *Bull. Epidémiol. Hebdomadaire* 10/93.

Kelkar, R., Gordon, S.M., Giri, N., Rao, K., Ramakrishnan, G., Sakia, T., Nair, C.N., Kurkure, P.A., Pai, S.K., Jarvis, W.R. and Advani, S.H. (1989). Epidemic iatrogenic *Acinetobacter* spp. *meningitis* following administration of intrathecal methotrexate. *J. Hosp. Infect.* 14, 233–243.

Kloos, W.E. and Musselwhite, M.S. (1975). Distribution and persistence of *Staphylococcus* and *Micrococcus* species and other aerobic bacteria on human skin. *Appl. Microbiol.* 30, 381–383.

Martinez, J.L., Cercenado, E., Baquero, F., Perez-Diaz, J.C. and Delgado-Tribarren, A. (1987). Incidence of aerobactin production in Gram-negative hospital isolates. *FEMS Microbiol. Lett.* 43, 351–353.

Musa, E.K., Desai, N. and Casewell, M.W. (1990). The survival of *Acinetobacter calcoaceticus* inoculated on fingertips and on formica. *J. Hosp. Infect.* 15, 219–227.

Nishimura, Y., Ino, T. and Iizuka, H. (1988). *Acinetobacter radioresistens* sp. nov. isolated from cotton and soil. *Int. J. Syst. Bacteriol.* 38, 209–211.

Obana, Y. (1986). Pathogenic significance of *Acinetobacter calcoaceticus*: analysis of experimental infection in mice. *Microbiol. Immunol.* 30, 645–657.

Obana, Y., Nishino, T. and Tanino, T. (1985). In-vitro and in-vivo activities of antimicrobial agents against *Acinetobacter calcoaceticus*. *J. Antimicrob. Chemother.* 15, 441–448.

Olafsson, M., Chen Lee, Y. and Abernathy, J.J. (1958). *Mima polymorpha* meningitis. Report of cases and review of the literature. *N. Engl. J. Med.* 258, 465–470.

Patterson, J.E., Vecchio, J., Pantelick, E.L., Farrel, P., Mazon, D., Zervos, M.J. and Hierholzer, W.J. (1991). Association of contaminated gloves with transmission of *Acinetobacter calcoaceticus* var. *anitratus* in an intensive care unit. *Am. J. Med.* 91, 479–483.

Peacock, J.E., Sorrell, L.A., Sottile, F.D., Price, L.E. and Rutala, W.A. (1988). Nosocomial respiratory tract colonization and infection with aminoglycoside-resistant *Acinetobacter calcoaceticus*: epidemiologic characteristics and clinical significance. *Infect. Control Hosp. Epidemiol.* 9, 302–308.

Picard, B. and Goullet, P. (1990). Epidemiological typing of *Acinetobacter* strains by esterase electrophoresis. *FEMS Microbiol. Lett.* 72, 229–234.

Pines, O. and Gutnick, D. (1984). Alternate hydrophobic sites on the cell surface of *Acinetobacter calcoaceticus* RAG-1. *FEMS Microbiol. Lett.* 22, 307–311.

Poh, C.L. and Loh, G.K. (1985). Enzymatic profile of clinical isolates of *Acinetobacter calcoaceticus*. *Med. Microbiol. Immunol.* 174, 29–33.

Pouëdras, P.S., Gras, S., Sire, J.M., Mesnard, R., Donnio, P.Y., and Avril, J.L. (1992). Esterase electrophoresis compared with biotyping for epidemiological typing of *Acinetobacter baumannii* strains. *FEMS Microbiol. Lett.* 96, 125–128.

Ramphal, R. and Kluge, R.M. (1979). *Acinetobacter calcoaceticus* var *anitratus*: an increasing nosocomial problem. *Am. J. Med. Sci.* 277, 57–66.

Retailliau, H.F., Hightower, A.W., Dixon, R.E. and Allen, J.R. (1979). *Acinetobacter calcoaceticus*: a nosocomial pathogen with an unusual seasonal pattern. *J. Infect. Dis.* 139, 371–375.

Reynolds, R.C. and Cluff, L.E. (1963). Infection of man with *Mimeae*. *Ann. Intern. Med.* 58, 759–762.

Robinson, R.G., Garrison, R.G. and Brown, R.W. (1964). Evaluation of the clinical significance of the genus *Herellea*. *Ann. Intern. Med.* 60, 19–27.

Rohea, H. and Guze, I.B. (1957). Infections due to *Bacterium anitratum*. *Arch. Intern. Med.* 100, 272–275.

Rosenberg, M., Bayer, E.A., Delarea, J. and Rosenberg, E. (1982). Role of thin fimbriae in adherence and growth of *Acinetobacter calcoaceticus* RAG-1 on hexadecane. *Appl. Environ. Microbiol.* 44, 929–937.

Rosenberg, E., Kaplan, N., Pines, O., Rosenberg, M. and Gutnick, D. (1983). Capsular polysaccharides interfere with adherence of *Acinetobacter calcoaceticus* to hydrocarbon. *FEMS Microbiol. Lett.* 17, 157–160.

Rosenthal, S.L. (1974). Sources of *Pseudomonas* and *Acinetobacter* species found in human culture materials. *Am. J. Clin. Pathol.* 62, 807–811.

Rosenthal, S. and Tager, I.B. (1975) Prevalence of Gram-negative rods in the normal pharyngeal flora. *Ann. Int. Med.* 83, 355–357.

Rudin, M.L., Michael, J.R. and Huxley, E.J. (1979). Community-acquired *Acinetobacter* pneumonia. *Am. J. Med.* 67, 39–43.

Sakata, H., Fujita, K., Maruyama, S., Kakehashi, H., Mori, Y. and Yoshioka, H. (1989). *Acinetobacter calcoaceticus* biovar *anitratus*, septicaemia in a neonatal intensive care unit: epidemiology and control. *J. Hosp. Infect.* 14, 15–22.

Seifert, H., Bouillon, B., Schulze, A. and Pulverer, G. (1994). Plasmid DNA profiles of *Acinetobacter baumannii*: clinical application in a complex endemic setting. *Infect. Control Hosp. Epidemiol.* 15, 520–524.

Sherertz, R.J. and Sullivan, M.L. (1985). An outbreak of infections with *Acinetobacter calcoaceticus* in burn patients: contamination of patients' mattresses. *J. Infect. Dis.* 151, 252–258.

Smego, R.A. (1985). Endemic nosocomial *Acinetobacter calcoaceticus* bacteremia: clinical significance, treatment, and prognosis. *Arch. Intern. Med.* 145, 2174–2179.

Smith, P.W. and Massanari, M. (1977). Room humidifiers as the source of *Acinetobacter* infections. *J. Am. Med. Assoc.* 237, 795–797.

Smith, A.W., Freeman, S., Minett, W.G. and Lambert, P.A. (1990). Characterization of a siderophore from *Acinetobacter calcoaceticus*. *FEMS Microbiol. Lett.* 70, 29–32.

Somerville, D.A. and Noble, W.C. (1970). A note on the Gram-negative bacilli of human skin. *Eur. J. Clin. Biol. Res.* 40, 669–670.

Stone, J.W. and Das, B.C. (1985). Investigation of an outbreak of infection with *Acinetobacter calcoaceticus* in a special care baby unit. *J. Hosp. Infect.* 6, 42–48.

Struelens, M.J., Carlier, E., Maes, N., Serruys, E., Quint, W.G.V. and van Belkum, A. (1993). Nosocomial colonization and infection with multiresistant *Acinetobacter baumannii*: outbreak delineation using DNA macrorestriction analysis and PCR-fingerprinting. *J. Hosp. Infect.* 25, 15–32.

Tankovic, J., Legrand, P., De Gatines, G., Chemineau, V., Brun-Buisson, C. and Duval, J. (1994). Characterization of a hospital outbreak of imipenem-resistant *Acinetobacter baumannii* by phenotypic and genotypic typing methods. *J. Clin. Microbiol.* 32, 2677–2681.

Taplin, D., Rebell, G. and Zaias, N. (1963). The human skin as a source of *Mima-Herellea* infections. *J. Am. Med. Assoc.* 186, 952–954.

Timsit, J.F., Garrait, V., Misset, B., Goldstein, F.W., Renaud, B. and Carlet, J. (1993). The digestive tract is a major site for *Acinetobacter baumannii* colonization in intensive care unit patients. *J. Infect. Dis.* 168, 1336–1337.

Tjernberg, I. and Ursing, J. (1989). Clinical strains of *Acinetobacter* classified by DNA-DNA hybridization. *Acta Pathol. Microbiol. Immunol. Scand.* 97, 596–605.

Traub, W. H. (1989). *Acinetobacter baumannii*, serotyping for delineation of outbreaks of nosocomial cross-infection. *J. Clin. Microbiol.* 27, 2713–2716.

Vandenbroucke-Grauls, C.M.J.E., Kerver, A.J.H., Rommes, J.H., Jansen R., den Dekker, C. and Verhoef, J. (1988). Endemic *Acinetobacter anitratus* in a surgical intensive care unit: mechanical ventilators as reservoir. *Eur. J. Clin. Microbiol. Infect. Dis.* 7, 485–489.

von Gravenitz, A. (1985). Ecology, clinical significance, and antimicrobial susceptibility of infrequently encountered glucose non-fermenting Gram-negative rods. In *Non-Fermentative Gram-Negative Rods. Laboratory Identification and Clinical Aspects*, pp. 181–232. G.L. Gilardi (Ed.). Marcel Dekker, New York.

Wands, J.R., Mann, R.B., Jackson, D. and Butler, T. (1973). Fatal community-acquired *Herellea* pneumonia in chronic renal disease. *Am. Rev. Respir. Dis.* 108, 964–966.

Wendt, C., Dietze, B. and Rüden, H. (1994). Survival of *Acinetobacter* species on dry surfaces. In *Abstr. 3rd Int. Symp. Biology of* Acinetobacter *(Edinburgh).*

Wise, K.A. and Tosolini, F.A. (1990). Epidemiological surveillance of *Acinetobacter* species. *J. Hosp. Infect.* 16, 319–329.

Chapter 5

Acinetobacter Infections — Overview of Clinical Features

Itzchak Levi and Ethan Rubinstein

CONTENTS

0-8493-9223-3/95/$0.00+$.50

5.1 INTRODUCTION

Acinetobacter spp. are Gram-negative coccobacilli that are distributed widely in the environment (Henriksen, 1973). Although *Acinetobacter* is an opportunistic organism of relatively low virulence, severe infections with these organisms do occur, especially in compromised patients after invasive diagnostic or therapeutic procedures.

Acinetobacter spp. can be isolated from virtually all samples of soil and water (Henriksen, 1973). However, the skin of patients and staff serves as a reservoir for *Acinetobacter* spp. in hospitals and is the likely source for most outbreaks of hospital infection (Al-Khoja and Darrell, 1979; see Chapter 4). Transmission is often attributed to common environmental sources and medical equipment, including respirators (Cunha et al., 1980) and solutions used for respiratory therapy (Hartstein et al., 1988; Cefai et al., 1990), intravenous catheters (Hamory, 1985; Seifert et al., 1993), pressure transducers (Beck-Sague et al., 1990), mattresses (Sherertz and Sullivan, 1985), or air and surfaces in the Intensive Care Unit (ICU) (Getchell-White et al., 1989). Even latex gloves used for universal precautions in an ICU have been associated with transmission of *Acinetobacter baumannii* (Patterson et al., 1991).

For many years it has been difficult to assess the clinical significance of *Acinetobacter* infections as the taxonomy of the genus has been changed frequently (see Chapter 2). Although most infections caused by these organisms are observed in very ill hospitalised patients, community-acquired infections have also been reported. *Acinetobacter* has caused bacteraemia, pneumonia, urinary tract infections, abscess, meningitis and wound infections. Endocardial, corneal and dental infections caused by this organism have also been described. In particular, with the advancement of diagnostic and therapeutic procedures in hospitals in the last two decades, *Acinetobacter* bacteraemia has increased in incidence.

Acinetobacter has emerged as a particularly important organism in ICUs (Bergogne-Bérézin and Joly-Guillou, 1991; see Chapter 4). *Acinetobacter* causes clinically significant infections in respiratory (Gerner-Smidt, 1987; Hartstein et al., 1988; Vandenbroucke-Grauls et al., 1988), neurosurgical (Siegman-Igra et al., 1993), neonatal (Ng et al., 1993) and burns (Sherertz and Sullivan, 1985) ICUs. The true frequency of nosocomial infection caused by *Acinetobacter* spp. is not easy to assess, in part because the isolation of *Acinetobacter* spp. from clinical specimens may not necessarily reflect infection, but rather result from colonisation (Struelens et al., 1993). Larson (1984) reviewed the prevalence of *Acinetobacter* nosocomial infection during a 10-year period (1971–1981) in a university hospital in the U.S. and found that *Acinetobacter* accounted for 1.4% of all nosocomial infections. Principal sites and types of infection included the respiratory tract, bacteraemia, peritoneum, urinary tract infection, surgical wounds, meningitis and skin or eye infection. In a more recent study of the prevalence of nosocomial *Acinetobacter* colonisation and infection in a university hospital (Struelens et al., 1993), hospitalisation in an ICU and previous administration of antibiotics was associated with *Acinetobacter* colonisation at various body sites in 3.2–10.8 per 1000 patients. *Acinetobacter* infection accounted for 0.3% of endemic nosocomial infections in critically ill patients, and for 1% of nosocomial bacteraemia hospital-wide (Struelens et al., 1993). These figures are in agreement with previous observations (Centers for Disease Control, 1987). Our experience at Chaim Sheba Medical Center, a tertiary medical centre in Israel, was similar in that the prevalence of *Acinetobacter* isolation over a 2-year period (1992–1993) was 1.4%.

One study has reported a seasonal incidence of *Acinetobacter* infection (Retailliau et al., 1979), with an increase in infection rates in late summer and early winter — a difference which could be related to temperature and humidity changes. However this observation has not yet been confirmed by other studies.

In summary, various factors predisposing to *Acinetobacter* infection have been identified over the years. Risk factors for nosocomial infection include recent major surgery, stay in ICUs, recent antimicrobial therapy, burns, malignant disease and/or other severe underlying disease.

5.2 RESPIRATORY TRACT INFECTIONS

5.2.1 Nosocomial Respiratory Tract Infections

Respiratory tract infection in ICU patients has been the most common infection associated with *Acinetobacter* (Bergogne-Bérézin and Joly-Guillou, 1991). Large outbreaks of *Acinetobacter* pneumonia have been described (Cunha et al., 1980; Bergogne-Bérézin et al., 1987; Hartstein et al., 1988). The patients affected most frequently are elderly patients with chronic disease, patients who have been supported with prolonged ventilatory assistance, and those who have received prolonged broad-spectrum antimicrobial therapy.

Acinetobacter respiratory tract infection is usually acquired via aspiration. Colonisation of the oropharynx with *Acinetobacter* spp. occurs in only about 7% of healthy people, and in these cases the organism burden is small (Rosenthal and Tager, 1975). The frequency of colonisation increases in hospitalised patients. Patients in ICUs are more likely to become colonised than those in general units (Johanson et al., 1972). Factors associated with higher rates of colonisation include the use of inhalation equipment, intubation and antimicrobial agents. Contaminated reusable ventilator support equipment may be a leading cause for colonisation with *Acinetobacter* of the sputum of ventilated patients. For example, an outbreak of respiratory infection and colonisation with *Acinetobacter* affecting 93 ventilated patients in all six of a hospital's ICUs over 10 months was traced to contaminated reusable ventilator circuits and resuscitation bags (Hartstein et al., 1988), with 18% of the equipment positive for *Acinetobacter*. Terminal ethylene oxide sterilisation of these devices was associated with prompt control of the outbreak. Plasmid DNA analysis of isolates from patients involved in the outbreak, from contaminated devices, and from the hands of personnel responsible for device disinfection, revealed two predominant plasmid profiles which represented two different isolates.

Once colonisation of the oropharyngeal mucosa occurs, the stage is set for contamination of the airways via aspiration. Diminished gag and cough reflexes, endotracheal or nasogastric intubation, and tracheostomy are all associated with repeated aspirations of infected oropharyngeal secretions (Levison and Kaye, 1985). *Acinetobacter* pneumonia has a crucial role in determining patient morbidity and mortality, with those patients that develop *Acinetobacter* pneumonia having a mortality rate of 71.4% compared to a mortality rate of 54.2% with pneumonia caused by other organisms, even after matching for severity of the underlying disease and other important demographic variables, date of admission and discharge from the ICU, and prior location in the hospital before admission to an ICU (Fagon et al., 1993).

The signs and symptoms of *Acinetobacter* pneumonia do not differ from those of other Gram-negative pneumonias, making a microbiological diagnosis essential. The most difficult task is to differentiate between colonisation and true infection when *Acinetobacter* is isolated from the sputum or tracheal aspirates. This is an especially difficult task in intubated patients receiving mechanical ventilation. A combination of signs, including fever, neutrophilia, sputum production and the appearance of new infiltrates on X-ray or CT scan, suggests respiratory tract infection (Johanson et al., 1972). Nevertheless, these signs may arise from atelectasis, pulmonary emboli, etc. Accurate and prompt microbiological investigation is therefore important and should be integrated with other physical findings to establish the correct diagnosis (Wunderink et al., 1992).

5.2.2 Community-Acquired Pneumonia

Community-acquired pneumonia caused by *Acinetobacter* is relatively rare. Sporadic cases have been reported from the U.S. (Goodhart et al., 1977), Papua, New Guinea (Barnes et al., 1988) and Australia (Anstey et al., 1992). Interestingly, such pneumonias are fulminant and carry a high mortality rate. Chronic pulmonary disease (CPD), diabetes mellitus, and tobacco and alcohol consumption appear to be major predisposing factors. A few cases have also been reported in persons without underlying disease. Rarely, community-acquired pneumonias caused by *Acinetobacter* occur in clusters, one example of which was in foundry workers who had been exposed to unacceptably high levels of total particulates, free silica and metallic dust (Cordes et al., 1981). Two of the three cases occurred in workers who had documented mixed dust pneumoconiosis.

The mean age of patients with community-acquired *Acinetobacter* pneumonia (51 years; range 41–70 years) is much younger than those suffering nosocomial pneumonia. Most (90%) of the reported cases are males, and 82% of cases occurred in the wet season. Most cases had at least two clinical risk factors, of which the most frequent were cigarette smoking and alcoholism. The most common clinical presentation was rapid onset of fever, dyspnoea, pleuritic chest pain, and a productive purulent cough yielding blood-stained sputum. Mean duration of symptoms was 2–3 days, and most patients experienced marked respiratory distress. Many patients presented with shock or severe hypoxaemia. The clinical course was fulminant with a mortality rate reaching 53–64%, with most casualties dying within the first 24 h. Shock at presentation and inappropriate initial antibiotic treatment were associated with poor prognosis. Initial chest roentgenograms showed classic unilateral consolidation in most patients, but did not help to distinguish between *Acinetobacter* and other organisms. Pleural effusions were rare, being present in only 2% of patients from the Northern Territory of Australia, but in 35% of patients from other areas. All reported *Acinetobacter* pulmonary infections appear to have been caused by *A. baumannii*.

Appropriate antimicrobial treatment of community-acquired pneumonia includes an aminoglycoside, an extended spectrum penicillin or third-generation cephalosporin, or imipenem. If the differential microbiological diagnosis includes *Acinetobacter* as a possible pathogen, either because of an underlying disorder or the presence of the patient in a high-risk area, then appropriate antibiotic therapy should be started pending laboratory results. Subsequent treatment can be tailored according to the results of sensitivity tests.

A more detailed account of respiratory tract infection with *Acinetobacter* spp. can be found in Chapter 6.

5.3 BACTERAEMIA AND SEPTICAEMIA

Little is known about *Acinetobacter* bacteraemia. A differentiation between contamination and true bacteraemia may at times be rather difficult. Tilley and Roberts (1994) suggested clinical criteria to differentiate between significant bacteraemia, transient bacteraemia, indeterminate bacteraemia and colonisation. According to these criteria, clinically significant bacteraemia is present when a bacteraemic episode is associated with relevant clinical signs, such as fever or leukocytosis that lasts for at least 8 h. Bacteraemia is considered transient when a bacteraemic episode lasts for a shorter period, with or without clinical findings, and is associated with instrumentation, surgical procedures or trauma. Bacteraemia is considered indeterminate when there is insufficient information to differentiate between positive blood cultures. Colonisation is considered when the organism is isolated from a single positive blood culture and there are no associated clinical signs to suggest an infection.

In most series, the commonest species causing significant bacteraemia is now identified as *A. baumannii*, with a rate of 0.13–1% (Smego, 1985). The highest rate of nosocomial bacteraemia occurs usually during the second week of hospitalisation. Immunocompromised patients comprise the largest group of patients with *Acinetobacter* bacteraemia, with malignant disease, trauma and burns being among the most common predisposing factors. In these patients the source of bacteraemia is often a respiratory tract infection. The prognosis of patients with malignant disease and burns is rather poor, but trauma patients have a better prognosis.

In general, the underlying disease determines the prognosis of the patient. Previous antibiotic treatment is associated with the selection of resistant strains (Tilley and Roberts, 1994). Some studies show that there is a correlation between vascular catheterisation and *Acinetobacter* infection (Rolston et al., 1985; Seifert et al., 1993). Beck-Sague et al. (1990) demonstrated an association between *A. baumannii* bacteraemia and the use of transducers for pressure monitoring. Rotation of the catheter insertion site every 2 days may reduce the risk. Hyperalimentation is another risk factor for bacteraemia (Beck-Sague et al., 1990). *Acinetobacter* may be found either as a single pathogen or as part of polymicrobial bacteraemia. Prompt attention to sterilisation techniques when handling equipment such as transducers may reduce the infection rate. Appropriate adherence to aseptic protocols is mandatory for vascular catheterisation.

5.4 INFECTIVE ENDOCARDITIS

Native valve infective endocarditis (IE) caused by *Acinetobacter* spp. is very rare. To date, only 16 cases have been reported in the English literature. Since *Acinetobacter* spp. are ubiquitous in the environment and are part of the indigenous flora of humans, many potential primary sources of infection exist, and about half of the reported cases of IE (both on native and on prosthetic valves) had no identifiable inciting event leading to endocarditis. For those cases in which identifying events could be found, 25% were associated with dental procedures, 16% followed open heart surgery, and single cases were reported in an intravenous drug user, following septic abortion, a stab wound, burns with use of intravenous catheters, upper respiratory tract infection with epistaxis, chronic otitis media, and a motor vehicle accident. Thus, any breach of the integument can result in seeding of a susceptible heart valve with *Acinetobacter*, leading to endocarditis.

Patients with native valve infectious endocarditis tend to be relatively young, with a mean age of 27 years (range 8–52 years). The underlying cardiac conditions are versatile, including rheumatic heart disease (30%), congenital heart disease (20%) and bicuspid aortic valvae (10%). About 40% of the reported cases (six patients) did not have any underlying cardiac disease. The aortic valve was affected most commonly (50%), with the mitral valve affected in 28% of cases, and both valves affected in 22% of cases.

Acinetobacter IE does not differ clinically from IE caused by other microorganisms, but there are wide variations in the presentation and clinical course of the disease. An interesting feature in most reported patients was splenomegaly caused by splenic embolisation with abscess formation. Among patients with prosthetic valves, *Acinetobacter* endocarditis appears to be less aggressive than the native valve form. More than 80% of patients with prosthetic valve IE survived, and none of these patients required surgical intervention. Empirical treatment should include a broad spectrum antibiotic. Imipenem should be administered initially, pending laboratory susceptibility results (Gradon et al., 1992).

5.5 WOUND AND BURN INFECTIONS

Soft tissue abscesses, burns (Green and Milling, 1983) and surgical wound infection, as well as animal bite wound infection (Auerbach and Morris, 1987; Martin et al., 1988) with *Acinetobacter* have been described. Wound infection with *Acinetobacter* may lead to bacteraemia. Its presence in an infected wound often accompanies additional Gram-negative and Gram-positive organisms (Amsel and Horrilleno, 1985). Therapeutic interventions require debridement with or without antimicrobial treatment.

There are a few reports in the literature describing *Acinetobacter* infections in burn patients. Data from the National Nosocomial Infections Study on 1,267 burn wound infections during a 4-year period, found *Acinetobacter* to be associated with <1% of these infections (Mayhall, 1981). Graber et al. (1962) found that 5% of burn patients developed bacteraemia caused by *Acinetobacter* spp., while Green and Milling (1983) reported on four patients who developed septicaemia over a 5-month period. It seems that patients colonised with *Acinetobacter* spp. have a high frequency of infections (68%) in which *Acinetobacter* is thought to be a pathogen (Sherertz and Sullivan, 1985). The most common sites of infections are usually the burn wound, urinary tract, bloodstream and lower respiratory tract. However, infections caused by *Acinetobacter* were less severe than infections caused by other agents. It seems that this organism is an uncommon problem in burns units, unless an outbreak occurs, providing that educated vigilance avoids the build-up of an environmental reservoir of these organisms (Sherertz and Sullivan, 1985).

5.6 URINARY TRACT INFECTIONS

5.6.1 Nosocomial Urinary Tract Infection

Nosocomial urinary tract infection is infrequently caused by *Acinetobacter* spp. It occurs most commonly in elderly debilitated patients, in patients confined to ICUs, and in patients with permanent indwelling urinary catheters. In one study (Pedraza et al., 1993), most patients were males (80%), probably reflecting the higher prevalence of indwelling urinary catheters in this population as a result of prostatic enlargement. Most (80%) of the patients in this study had urinary catheters.

The clinical manifestations of nosocomial *Acinetobacter* UTI do not differ from UTI caused by other organisms. However, not every isolation of *Acinetobacter* from the urinary tract means infection, especially in the case of patients with an indwelling urinary catheter (Hoffmann et al., 1982).

5.6.2 Community-Acquired Urinary Tract Infection

The possible medical significance of *Acinetobacter* spp. in the human genito-urinary system was first brought to attention when De Bord (1939) described a new tribe called *Mimeae* (*Mima* and *Herellea*). Several reviews have commented upon the presence of these organisms in urinary cultures. In general practice, *Acinetobacter* spp. was found in 4% of patients with significant bacteriuria (Thamsborg et al., 1980). Of 2,314 individuals who participated in a survey, 12% had significant bacteriuria, and *Acinetobacter* was isolated from 14% of these (Hoffmann et al., 1982). Robinson et al. (1964) reported that 24 males had urine cultures that were positive for *Acinetobacter* spp., but in only three cases was the finding significant. Factors predisposing to community-acquired urinary tract infection caused by *Acinetobacter* do not seem to differ from those predisposing to infections caused by other bacteria. Such factors are chronic pyelonephritis, diabetes mellitus and malformations of the urinary tract.

5.7 MENINGITIS

Acinetobacter is a rare cause of meningitis. Sporadic cases have been reported following neurosurgical procedures or head trauma. Until 1967, there were about 60 reports of *Acinetobacter* meningitis, most of which were community acquired. However, since 1979, the vast majority of cases have been nosocomial infections, and almost all were caused by *A. baumannii*. As an exception, an outbreak of *Acinetobacter* meningitis was described in a group of children with leukaemia by Kelkar et al. (1989) following the administration of intrathecal methotrexate. The outbreak was caused by the use of inappropriately sterilised needles. Eight of the 20 children who received intrathecal methotrexate returned within 2–19 h of treatment with signs and symptoms of acute meningeal irritation. *Acinetobacter* was isolated from the CSF of five of these patients, as well as from the methotrexate solution. Three of the children died as a result of meningitis and five recovered.

In general, in the last decade *Acinetobacter* meningitis has been a nosocomial infection. Most patients have been adults, predominantly male, and had undergone lumbar punctures, myelography, ventriculography, or other neurosurgical procedures, although one patient had post-traumatic otorrhoea without intervention. Mortality rates from different series range from 20–27%. Risk factors include the presence of a continuous connection between the ventricles and the external environment, a ventriculostomy, or a CSF fistula. In addition, the presence of an indwelling ventricular catheter for more than 5 days was found to be an important risk factor. Another important predisposing factor was the heavy use of antimicrobial agents in the neurosurgical ICU. One outbreak subsided spontaneously only when the selective pressure of antibiotics was reduced (Siegman-Igra et al., 1993). Only a very few cases of community-acquired *Acinetobacter* meningitis have been described in recent years.

5.8 PERITONITIS IN PATIENTS ON CONTINUOUS AMBULATORY PERITONEAL DIALYSIS

Acinetobacter spp. are important organisms causing peritonitis in patients undergoing continuous ambulatory peritoneal dialysis (CAPD), with 14% of dialysis-associated peritonitis episodes being caused by these organisms. Most episodes are caused by *A. baumannii*. Technique failure and diabetes mellitus are the most common underlying risk factors. Mean duration of risk factors before the onset of peritonitis ranged from 2–13 months. The most common manifestations were abdominal pain or tenderness and cloudy dialysate, but only a minority of patients had fever. Most patients respond to antibiotic therapy without the need to interrupt CAPD (Galvao et al., 1989; Lye et al., 1991; Valdez et al., 1991).

5.9 NEONATAL INFECTION AND SEPSIS

Although *Acinetobacter* is one of the organisms responsible for nosocomial infection in ICUs, neurosurgery and burns units, there are only a few reports of *Acinetobacter* infection in neonates. In one report from Japan, 19 neonates with *Acinetobacter* septicaemia were described in the neonatal ICU over a period of 30 months. All cases were of late-onset type septicaemia in infants hospitalised for long periods, with a mortality rate of 11%. *Acinetobacter* was isolated from the hands of the staff, from sinks, and from hospitalised infants (Sakata et al., 1989). Although faecal or rectal colonisation with *Acinetobacter* has been reported to occur in less than 9% of patients (Fujita et al., 1982), the faecal carriage rate of infants in this particular study was considerably higher (40%), and the intestinal flora of the neonates was considered to be the reservoir of the organism in this outbreak. Before the introduction of control measures, 72% of 72 hospitalised low-birth-weight (less than

2000 g) infants were colonised, and 19% had septicaemia with *Acinetobacter*. The predisposing risk factors for septicaemia were low birth weight, previous antibiotic therapy, mechanical ventilation and the presence of neonatal convulsions. The introduction of strict aseptic techniques, especially hand washing and separation of colonised from non-colonised infants, as well as restriction of antibiotic use, were effective in reducing the colonisation rate and putting an end to the outbreak.

In addition to *A. baumannii*, *A. lwoffii* can also cause neonatal septicaemia. Ng et al. (1993) described an outbreak of *A. lwoffii* septicaemia in a neonatal intensive care unit, which was confined to seven babies receiving parental nutrition. All babies had severe clinical symptoms and signs of septic shock. Five required ventilatory support. All babies recovered uneventfully after appropriate antibiotic therapy. Interestingly, thrombocytopenia was a prominent and consistent feature, which appeared to be most severe on days 3–5 of the illness.

In a more recent study from Israel, Regev et al. (1993) described nine cases of *Acinetobacter* sepsis in a neonatal intensive care unit over a period of 31 months. The clinical course was fulminant in four babies that subsequently died. Thus, *Acinetobacter* should be added to the list of organisms capable of causing severe nosocomial infection in neonatal ICUs.

5.10 OTHER *Acinetobacter* INFECTIONS

Acinetobacter cholangitis and septic complications following percutaneous transhepatic cholangiogram (PTC) and percutaneous biliary drainage (PBD) occur primarily among elderly patients with obstructive jaundice caused by malignant disease or choledocholithiasis. In one study, 13.5% of patients developed sepsis, with the most common isolates being *Enterobacter cloacae* and *A. baumannii* (Sacks-Berg et al., 1992). Other more rare case reports include typhilitis after autologous bone marrow transplantation (Nagler et al., 1992), and osteomyelitis and extremities infections following injury (Dietz et al., 1988; Martin et al., 1988). Eye infections following trauma have also been described (Mark and Gaynon, 1983). Both *A. baumannii* and *A. lwoffii* (Melki and Sramek, 1992) have been implicated in trauma-induced endophthalmitis. Penetrating keratoplasty (Zabel et al., 1989) and the fitting of contact lens (Barre and Cook, 1984) have also caused ophthalmic infections with *Acinetobacter*.

Acinetobacter is isolated frequently from polymicrobial infections, and such infections are reported to have resulted in severe disease. Several studies of patients with suspected pneumonia have shown that

Acinetobacter is associated frequently with *Staphylococcus aureus* and/or *Pseudomonas aeruginosa*. Among patients with nosocomial pneumonia caused by *Acinetobacter* spp., mixed infections were observed in 3/3 patients (Marquette et al., 1993); 4/5 patients (Montravers et al., 1993); 2/3 patients (Torres et al., 1993); and 4/8 patients (Joly-Guillou et al., 1993). Experimentally, Obana (1986) has shown that mixed infections combining certain other species of bacteria with *Acinetobacter* were more virulent than infections with a single organism alone. The slime produced by *Acinetobacter* spp. was considered to be the main factor responsible for the enhancement of virulence in mixed infections.

5.11 SUMMARY

Acinetobacter has increased in importance in the last decade, especially in its role as a nosocomial pathogen. Patients in ICUs are especially prone to infection with multiresistant isolates of *A. baumannii*. Since *Acinetobacter* is a ubiquitous organism in the hospital environment, it is often difficult to distinguish colonisation from true infection. Signs of infection, combined with isolation of the organism from the site of infection should prompt the physician to start antimicrobial treatment. A rise in the incidence of *Acinetobacter* infection in a hospital or unit should prompt the infectious disease team to investigate the epidemiology of the possible outbreak, which can usually be attributed to spread by staff or equipment in the environment.

REFERENCES

Al-Khoja, M.S. and Darrell, J.H. (1979). *Acinetobacter* and *Moraxella* species occurring in blood cultures. *J. Clin. Pathol.* 32, 497–499.

Amsel, M.B. and Horrilleno, E. (1985). Synergistic necrotizing fascitis: a case of polymicrobial infection with *Acinetobacter calcoaceticus*. *Curr. Surg.* 42, 370–372.

Anstey, M., Currie, B.J. and Withnall, K.M. (1992). Community acquired pneumonia in the northern territory of Australia. *Clin. Infect. Dis.* 14, 83–91.

Auerbach, P.S. and Morris, J. (1987). *Acinetobacter calcoaceticus* infection following a dog bite. *J. Emergency Med.* 5, 363–366.

Barnes, D.J., Naraqi, S. and Igo, J.D. (1988). Community acquired *Acinetobacter* pneumonia in adults in Papua New Guinea. *Rev. Infect. Dis.* 10, 636–639.

Barre, M.E. and Cook, M.L. (1984). Microbial factors in contact lens fitting. *Am. J. Optometry Physiol. Opt.* 61, 389–396.

Beck-Sague, C.M., Jarvis, W.R., Brook, J.H., Culver, D.H., Potts, A., Gay, E., Shotts, B.W., Hill, B., Anderson, R.L. and Weinstein, M.P. (1990). Epidemic bacteremia due to *Acinetobacter baumannii* in five intensive care units. *Am. J. Epidemiol.* 132, 723–733.

Bergogne-Bérézin, E. and Joly-Guillou, M.-L. (1991). Hospital infection with *Acinetobacter* spp.: an increasing problem. *J. Hosp. Infect.* 18 (Suppl. A), 250–255.

Bergogne-Bérézin, E., Joly-Guillou, M.-L. and Vieu, J.F. (1987). Epidemiology of nosocomial infections due to *Acinetobacter calcoaceticus*. *J. Hosp. Infect.* 10, 105–113.

Cefai, C., Richards, J., Gould, F.K. and McPeake, P. (1990). *Acinetobacter* respiratory tract infection resulting from incomplete disinfection of ventilatory equipment. *J. Hosp. Infect.* 15, 177–182.

Centers for Disease Control (1987). Nosocomial infection surveillance 1984. *CDC Summaries* 35, 17SS–29SS.

Cordes, L.G., Brink, E.W., Checko, P.J., Lentnek, A., Lyons, R.W., Hayes, P.S., Wu, T.C., Tharr, D.G. and Fraser, D.W. (1981). A cluster of *Acinetobacter* pneumonia in foundry workers. *Ann. Intern. Med.* 95, 688–693.

Cunha, B.A., Klimek, J.J., Gracewski, J., McLaughlin, J.C. and Quintiliani, R. (1980). A common source outbreak of *Acinetobacter* infections traced to Wright respirometers. *Postgrad. Med. J.* 56, 169–172.

DeBord, G.G. (1939). Organisms invalidating the diagnosis of *gonorrheae* by the smear method. *J. Bacteriol.* 38, 119–120.

Dietz, J.W., Goodrich, J.A. and Brown, W.B. (1988). *Acinetobacter calcoaceticus* foot infection secondary to high pressure injection injury: a case report. *Foot Ankle* 8, 216–222.

Fagon, J.Y., Chastre, J., Hance, A., Montravers, P., Novara, A. and Gibert, C. (1993). Nosocomial pneumonia in ventilated patients — a cohort study evaluating attributable mortality and hospital stay. *Am. J. Med.* 94, 281–287.

Fujita, K., Lilly, H.A. and Ayliffe, G.A.J. (1982). Spread of resistant Gram-negative bacilli in a burns unit. *J. Hosp. Infect.* 3, 29–37.

Galvao, C., Swartz, R., Rocher, L., Reynolds, J., Starmann, B. and Wilson, D. (1989). *Acinetobacter* peritonitis during chronic peritoneal dialysis. *Am. J. Kidney Dis.* 14, 101–104.

Gerner-Smidt, P. (1987). Endemic occurrence of *Acinetobacter calcoaceticus* biovar *anitratus* in an intensive care unit. *J. Hosp. Infect.* 10, 265–272.

Getchell-White, S.I., Donowitz, L.G. and Groschel, D.H.M. (1989). The inanimate environment of an intensive care unit as a potential source of nosocomial bacteria: evidence for long survival of *Acinetobacter calcoaceticus*. *Infect. Control Hosp. Epidemiol.* 10, 402–407.

Goodhart, G.L., Abrutyn, E., Watson, R., Root, R.K. and Egert, J. (1977). Community acquired *Acinetobacter calcoaceticus* var *anitratus* pneumonia. *J. Am. Med. Assoc.* 238, 1516–1518.

Graber, C.D., Rabin, E.R., Mason, A.D. and Vogel, E. (1962). Increasing incidence of nosocomial *Herellea vaginicola* infections in burned patients. *Surg. Gynecol. Obstet.* 114, 109–112.

Gradon, D.J., Chapnick, E.K. and Lutwick, L.I. (1992). Infective endocarditis due to *Acinetobacter*: case report and review. *Clin. Infect. Dis.* 14, 1145–1148.

Green, A.R. and Milling, M.A.P. (1983). Infection with *Acinetobacter* in a burns unit. *Burns* 9, 292–294.

Hamory, B.H. (1985). Nosocomial bloodstream and intravascular device related infection. In *Prevention and Control of Nosocomial Infections*, pp. 283–319. R.P. Wenzel (Ed.). Williams & Wilkins, Baltimore.

Hartstein, A.I., Rashad, A.L., Liebler, J.M., Actis, L.A., Freeman, J., Rourke, J.W., Stibolt, T.B., Tomarsky, M.E., Ellis, G.R. and Croser, J.H. (1988). Multiple intensive care unit outbreak of *Acinetobacter calcoaceticus* subspecies *anitratus* respiratory infection and colonisation associated with contaminated, reusable ventilator circuits and resuscitation bags. *Am. J. Med.* 85, 624–631.

Henriksen, S.D. (1973). *Moraxella, Acinetobacter*, and the *Mimiae*. *Bacteriol. Rev.* 37, 522–561.
Hoffmann, S., Mabeck, C.E. and Vejsgaard, R. (1982). Bacteriuria caused by *Acinetobacter calcoaceticus* biovars in a normal population and in general practice. *J. Clin. Microbiol.* 16, 443–451.
Johanson, W.G., Pierce, A.K., Sanford, J.P. and Thomas, G.D. (1972). Nosocomial respiratory infections with Gram-negative bacilli: the significance of colonization of the respiratory tract. *Ann. Intern. Med.* 77, 701–706.
Joly-Guillou, M.-L., Sollet, J.P., Varache, C. and Bergogne-Bérézin, E. (1993). Infections dues à *Acinetobacter baumannii*. *Méd. Malad. Infect.* 23S, 67–72.
Kelkar, R., Gordon, S.M., Giri, N., Rao, K., Ramakrishnan, G., Saikia, T., Nair, C.N., Kurkure, P.A., Pai, S.K., Jarvis, W.R. and Advani, S.H. (1989). Epidemic iatrogenic *Acinetobacter* spp. meningitis following administration of intrathecal methotrexate. *J. Hosp. Infect.* 14, 233–243.
Larson, E. (1984). A decade of nosocomial *Acinetobacter*. *Am. J. Infect. Control* 12, 14–18.
Levison, M.E. and Kaye, D. (1985). Pneumonia caused by Gram-negative bacilli: an overview. *Rev. Infect. Dis.* 7 (Suppl. 4), S656–S664.
Lye, W.C., Lee, E.J. and Ang, K.K. (1991). *Acinetobacter* peritonitis in patients on CAPD; characteristics and outcome. *Adv. Peritoneal Dialysis* 7, 176–179.
Mayhall, C.G. (1981). Infections in burn patients. In *CRC Handbook of Hospital Infections*, pp. 317–339. R.P. Wenzel (Ed.). CRC Press, Boca Raton, FL.
Mark, D.B. and Gaynon, M.W. (1983). Trauma induced endophthalmitis caused by *Acinetobacter anitratus*. *Br. J. Ophthalmol.* 67, 124–126.
Marquette, C.H., Georges, H., Wallet, F., Ramon, P., Saulnier, F., Neviere, R., Mathieu, D., Rime, A. and Tonel, A.B. (1993). Diagnostic efficiency of endotracheal aspirates with quantitative cultures in intubated patients with suspected pneumonia. *Am. Rev. Resp. Dis.* 148, 138–144.
Martin, R.W., Martin, D.L. and Levy, C.S. (1988). *Acinetobacter* osteomyelitis from a hamster bite. *Pediatr. Infect. Dis. J.* 7, 364–365.
Melki, T.S. and Sramek, S.J. (1992). Trauma induced *Acinetobacter lwoffii* endophthalmitis. *Am. J. Ophthalmol.* 113, 598–599.
Montravers, P., Fagon, J.Y., Chastre, J., Lecso, M., Combret, M.C., Trouiler, J.L. and Gibert, C. (1993). Follow up protected specimen brushes to assess treatment in nosocomial pneumonia. *Am. Rev. Respir. Dis.* 147, 38–44.
Nagler, A., Pavel, L., Naparstek, E., Muggiasullam, M. and Slavin, S. (1992). Typhilitis occurring in autologous bone marrow transplantation. *Bone Marrow Transplant.* 9, 63–64.
Ng, P.C., Herrington, R.A., Beane, C.A., Ghoneim, A.T.M. and Dear, P.R.F. (1993). Nosocomial colonisation and infection with multiresistant *Acinetobacter baumannii*: outbreak delineation using DNA macrorestriction analysis and PCR-fingerprinting. *J. Hosp. Infect.* 25, 15–32.
Obana, Y. (1986). Pathogenic significance of *Acinetobacter calcoaceticus:* analysis of experimental infection in mice. *Microbiol. Immunol.* 30, 645–657.
Patterson, J.E., Vecchio, J., Pantelick, E.L., Farrel, P., Mazon, D., Zervos, M.J. and Hierholzer, W.J. (1991). Association of contaminated gloves with transmission of *Acinetobacter calcoaceticus* var. *anitratus* in an intensive care unit. *Am. J. Med.* 91, 479–483.
Pedraza, F., Andreu, A., Saune, M., Moreno, A., Ramirez, L. and Garcia, L. (1993). A urinary outbreak of *Acinetobacter baumannii* in a spinal cord injury unit. *Ann. Med. Int.* 10, 55–58.
Regev, R., Dolfin, T., Zelig, T., Givoni, S. and Wolach, B. (1993). *Acinetobacter* septicemia: a threat to neonates? Special aspects in a neonatal intensive care unit. *Infection* 21, 394–396.

Retailliau, H.F., Hightower, A.W., Dixon, R.E. and Allen, J.R. (1979). *Acinetobacter calcoaceticus:* a nosocomial pathogen with an unusual pattern. *J. Infect. Dis.* 139, 371–375.

Robinson, R.G., Garrbon, R.G. and Brown, R.W. (1964). Evaluation of the clinical significance of the genus *Herellea. Ann. Intern. Med.* 60, 19–27.

Rolston, K., Guan, Z., Bodey, G.P. and Elting, L. (1985). *Acinetobacter calcoaceticus* septicemia in patients with cancer. *South. Med. J.* 78, 647–651.

Rosenthal, S. and Tager, T.B. (1975). Prevalence of Gram-negative rods in the normal pharyngeal flora. *Ann. Intern. Med.* 83, 355–357.

Sacks-Berg, A., Calubiran, O.V., Epstein, H.Y. and Cunha, B.A. (1992). Sepsis associated with transhepatic cholangiography. *J. Hosp. Infect.* 20, 43–50.

Sakata, H., Fujita, K., Maruyama, S., Kakehashi, H., Mori, Y. and Yoshioka, H. (1989). *Acinetobacter calcoaceticus* biovar *anitratus* septicaemia in a neonatal intensive care unit: epidemiology and control. *J. Hosp. Infect.* 14, 15–22.

Seifert, H., Strate, A., Schulze, A. and Pulverer, G. (1993). Vascular catheter related bloodstream infection due to *Acinetobacter johnsonii* (formerly *Acinetobacter calcoaceticus* var. *lwoffi*): report of 13 cases. *Clin. Infect. Dis.* 17, 632–636.

Sherertz, R.J. and Sullivan, M.L. (1985). An outbreak of infections with *Acinetobacter calcoaceticus* in burn patients: contamination of patients' mattresses. *J. Infect. Dis.* 151, 252–258.

Siegman-Igra, Y., Bar Yosef, S., Gorea, A. and Avram, J. (1993). Nosocomial *Acinetobacter* meningitis secondary to invasive procedures: report of 25 cases and review. *Clin. Infect. Dis.* 17, 843–849.

Smego, R.A. (1985). Endemic nosocomial *Acinetobacter calcoaceticus* bacteremia. Clinical significance, treatment and prognosis. *Arch. Intern. Med.* 145, 2174–2179.

Struelens, M.J., Carlier, E., Maes, N., Serruys, E., Quint, W.G.V. and van Belkum, A. (1993). Nosocomial colonisation and infection with multiresistant *Acinetobacter baumannii:* outbreak delineation using DNA macrorestriction analysis and PCR-fingerprinting. *J. Hosp. Infect.* 25, 15–32.

Thamsborg, G.M., Bakslev, I., Mabeck, C.E. and Vejkgaard, R. (1980). Urinvejsinfektioner i almen praksis I kmimisk billede og bakterieflora. *Ugeskr Laeg.* 142, 1657–1660.

Tilley, P.A.G. and Roberts, F.J. (1994). Bacteremia with *Acinetobacter* species: risk factors and prognosis in different clinical settings. *Clin. Infect. Dis.* 18, 896–900.

Torres, A., Martos, A., de la Bellacasa, J.P., Ferrer, M., El-Ebiary, M., Gonzalez, J., Gené, A. and Rodriguez-Roisin, R. (1993). Specificity of endotracheal aspiration, protected specimen brush, and bronchoalveolar lavage in mechanically ventilated patients. *Am. Rev. Respir. Dis.* 147, 952–957.

Valdez, J.M., Asperilla, M.O. and Smego, R.A. (1991). *Acinetobacter* peritonitis in patients receiving continuous ambulatory peritoneal dialysis. *South. Med. J.* 84, 607–610.

Vandenbroucke-Grauls, C.M.J.E., Kerver, A.J.H., Rommes, J.H., Jansen, R., den Dekker, C. and Verhoef, J. (1988). Endemic *Acinetobacter anitratus* in a surgical intensive care unit: mechanical ventilators as reservoir. *Eur. J. Clin. Microbiol. Infect. Dis.* 7, 485–489.

Wunderink, R.G., Woldenberg, L.S., Zeiss, J., Day, C.M., Ciemins, J. and Lacher, D.A. (1992). The radiologic diagnosis of autopsy proven ventilator associated pneumonia. *Chest* 101, 458–463.

Zabel, R.W., Winegarden, T., Holland, E.J. and Doughman, D.J. (1989). *Acinetobacter* corneal ulcer after penetrating keratoplasty. *Am. J. Ophthalmol.* 107, 677–678.

Chapter 6

NOSOCOMIAL PNEUMONIA CAUSED BY *Acinetobacter* spp.

Jean Chastre, Jean-Louis Trouillet, Albert Vuagnat, and Marie-Laure Joly-Guillou

CONTENTS

0-8493-9223-3/95/$0.00+$.50

6.1 INTRODUCTION

Until 1970, clinicians had generally ignored evidence that *Acinetobacter* spp. could be the microbiological cause of pneumonia, and considered these organisms to be of relatively low virulence in humans. In recent years, however, *Acinetobacter* spp. have assumed increasing importance in nosocomial infections, particularly in the intensive care setting (see Chapter 4). This increased importance results, in part, from the dramatic differences in the profile of patients admitted and cared for in intensive care units (ICUs) over the past 20 years; patients are now older with more severe underlying disease, and medical care is more likely to include invasive devices, major surgery, prolonged antimicrobial treatment and/or immunosuppressive therapy, all of which increase susceptibility to bacterial infection and subsequent host mortality.

Despite the increasing role played by *Acinetobacter* spp. in nosocomial pneumonia in ICU patients, relatively few studies have been published, and these investigations have usually involved small numbers of patients with infection caused by this organism. Understanding the epidemiology of these infections has so far been difficult, partly because of problems in distinguishing patients who are only colonised from patients with true pulmonary infection, and partly because of the lack of a reliable typing system to evaluate the relatedness of isolates collected during ICU outbreaks. This chapter reviews published studies on the epidemiology and clinical characteristics of nosocomial pneumonia caused by *Acinetobacter* spp. in critically ill patients hospitalised in ICUs and presents our experience with this infection.

6.2 INCIDENCE OF NOSOCOMIAL *Acinetobacter* PNEUMONIA

As described in Chapters 4 and 5, *Acinetobacter baumannii* is now recognised to be the *Acinetobacter* species associated with the majority of human infections caused by this genus, including nosocomial outbreaks of pulmonary infection (Allen and Hartman, 1990; Bergogne-Bérézin and Joly-Guillou, 1991). While nosocomial infections caused by *Acinetobacter* spp. remain relatively uncommon, accounting for <1% of the burden of endemic infections in most hospitals (Centers for Disease Control, 1987), a number of outbreaks of nosocomial pulmonary infection have been described in ICUs, and the role played by these organisms in ventilator-associated pneumonia is obviously increasing (Buxton et al., 1978; Castle et al., 1978; Cunha et al., 1980; Stone and Das, 1985; Hartstein et al., 1988; Vandenbroucke-Grauls et al., 1988; Cefai et al.,

TABLE 6.1

Pathogens Most Frequently Reported Associated with Nosocomial Pneumonia in Patients Enrolled in the NNIS Registry from 1990–1992

(n = 8,891 isolates) Pathogen	%	Rank in 1992	Rank in 1988	Rank in 1984
S. aureus	20	1	2	2
P. aeruginosa	16	2	1	1
Enterobacter spp.	11	3	3	4
K. pneumoniae	7	4	4	3
H. influenzae	5	5	6	—
C. albicans	5	6	8	8
Acinetobacter spp.	4	7	10	—
E. coli	4	8	5	5

1990; Bergogne-Bérézin and Joly-Guillou, 1991). Regardless of the bacteriological method used to define the cause of pneumonia precisely, several studies have reported that about 3–5% of nosocomial pneumonias are caused by *Acinetobacter* (Centers for Disease Control, 1987; Craven et al., 1990). In patients with pneumonia enrolled in the National Nosocomial Infection Study (Schaberg et al., 1991), this organism accounted for 4% of the total number of pulmonary infections, and was ranked in seventh position in the list of pathogens identified most frequently in the survey, while it was not included in this list in 1984, and was in tenth position in 1988 (Table 6.1).

Accurate data on the epidemiology of *A. baumannii* nosocomial pneumonia in critically ill patients cared for in the ICU are limited by the lack of a gold standard for diagnosis, especially in patients undergoing mechanical ventilation (Craven et al., 1990; Chastre and Fagon, 1994). Consequently, nearly all previous epidemiological investigations in this setting have relied solely on clinical diagnostic criteria, and have therefore probably included patients who did not have pneumonia. Furthermore, in most of these studies, tracheal secretions or sputum were used as the major specimen source for microbiological analysis, despite the fact that the upper respiratory tract of many of these patients may be colonised by *Acinetobacter* spp. in the absence of true pulmonary parenchymal infection. In addition, many epidemiological studies have been conducted in a clinical setting corresponding to a major outbreak; thus, probably only a limited portion of the clinical spectrum of the disease was included since *A. baumannii* is now endemic in many ICUs and hospitals (Bergogne-Bérézin and Joly-Guillou, 1991). Finally, only a few studies have used an adequate system to differentiate isolates of *Acinetobacter* clearly from one another, leading to the risk of concluding, falsely, that a large outbreak was occurring, while in fact several different strains were present (Allardet-Servent et al., 1989; Bouvet, 1991).

Since it may be difficult to determine whether or not pneumonia has developed in critically ill patients, especially if they are undergoing mechanical ventilation, a number of specialised microbiological assays and invasive techniques for obtaining specimens have been described as being potentially useful for improving the diagnostic accuracy for nosocomial pneumonia (Chastre and Fagon, 1994). Of these various techniques, two have been suggested to be of particular value in establishing a specific diagnosis of pneumonia in hospitalised ventilator-dependent patients: first, the use of a bronchoscopic double-lumen catheter with a protected specimen brush (PSB) to collect uncontaminated culture specimens directly from affected areas in the lower respiratory tract for quantitative cultures; and second, the use of bronchoalveolar lavage (BAL), since this technique is a safe and practical method for obtaining cells and secretions from a large area of the lung that can be examined microscopically immediately after the procedure and that are also suitable for culture by quantitative techniques. Many investigators have now confirmed that secretions obtained by these two techniques, combined with quantitative culture, offer a sensitive and specific approach for establishing the organisms responsible for pneumonia in critically ill patients, and for differentiating between colonisation of the upper respiratory tract and distal lung infection (Fagon et al., 1989a; Cook et al., 1991).

The importance of using such bronchoscopic techniques in epidemiological studies of nosocomial pneumonia caused by *Acinetobacter* spp. and other non-fermenting Gram-negative bacteria is illustrated by a study performed by our group at Bichat Hospital in Paris. This study evaluated the crude mortality observed in a large group of 52 consecutive mechanically ventilated patients who had developed nosocomial pneumonia caused by *Pseudomonas* or *Acinetobacter* spp., according to results obtained by specific bronchoscopic techniques, compared with a control group of 33 patients who had only tracheobronchitis caused by the same microorganisms (Fagon et al., 1989b). The overall mortality of patients without pneumonia who received no specific antimicrobial treatment was 28%, compared with 75% in patients who developed true pneumonia caused by *Acinetobacter* or *Pseudomonas* spp. (Figure 6.1). These results underline the necessity in epidemiological studies of clearly distinguishing patients with only proximal colonisation from patients with true infection.

Using specific diagnostic techniques, several recent investigations have demonstrated the increasing role played by *A. baumannii* in nosocomial pneumonia for the subset of ICU patients requiring mechanical ventilation. In the studies of Fagon et al. (1989b) and Torres et

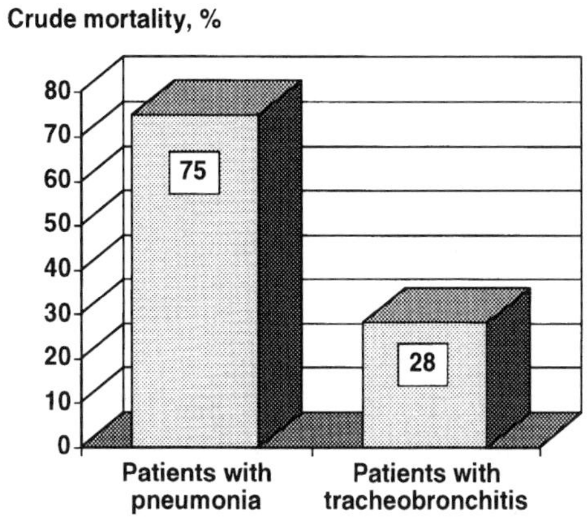

FIGURE 6.1
Crude mortality observed in 52 consecutive mechanically ventilated patients with nosocomial pneumonia caused by *Acinetobacter* or *Pseudomonas* spp., compared with a control group of 33 patients with tracheobronchitis caused by the same organisms (data taken from Fagon et al., 1989b).

al. (1990), in which only mechanically ventilated patients were included and bacteriological studies were restricted to uncontaminated specimens obtained by bronchoscopic techniques, 15 and 24% of all episodes of pneumonia, respectively, included at least one *Acinetobacter* spp. (Table 6.2). These data suggest clearly that nosocomial pneumonia caused by *A. baumannii* is now emerging as a prominent complication of mechanical ventilation. Interestingly, this increased incidence has been observed despite many major advances in techniques in the management of ventilator-dependent patients and the routine use of effective procedures to disinfect respiratory equipment.

In order to try to better define the epidemiological characteristics of ventilator-associated pneumonia caused by *A. baumannii*, Trouillet et al. (1995) studied 159 consecutive patients who received mechanical ventilation for >72 h in their medical-surgical ICU over a 13-month period. Fibreoptic bronchoscopy with protected specimen brush (PSB) and bronchoalveolar lavage (BAL) was performed on each patient suspected of having pneumonia because of the presence of a new pulmonary infiltrate, fever and purulent tracheal secretions, but the diagnosis of pulmonary infection was retained only if PSB and/or BAL specimens grew $\geq 10^3$ and $\geq 10^4$ cfu of at least one microorganism/ml, respectively. By the criteria used in this study, *A. baumannii* nosocomial pneumonia occurred in 19 (12%) of the 159 study patients, and played a role in 27% of ventilator-associated pneumonia cases diagnosed during this period of time. It is noteworthy that this very high incidence of nosocomial

TABLE 6.2

Organisms Recovered from Ventilator-Associated Pneumonia

	Fagon et al. (1989b) (n = 52) %	Torres et al. (1990) (n = 25) %
Gram-negative bacteria		
P. aeruginosa	31	28
Acinetobacter spp.	15	24
Proteus spp.	15	0
Moraxella catarrhalis	10	0
Haemophilus spp.	10	0
E. coli	8	12
Miscellaneous	10	28
Gram-positive bacteria		
S. aureus	33	20
Miscellaneous	29	24
Anaerobes	2	4
Polymicrobial flora	40	40

Acinetobacter infection was observed in the absence of any increase in the overall rate of *Acinetobacter* isolation in clinical samples collected during the study period when compared to the previous months. By an actuarial method, the cumulative risk of pneumonia caused by this microorganism was estimated to be 3.4% at 10 days, 20% at 20 days and 48% at 30 days after the onset of mechanical ventilation, and was therefore increasing nonlinearly with the duration of ventilation. Interestingly, this risk was similar to the risk of *P. aeruginosa* ventilator-associated pneumonia (Figure 6.2), but was lower than the risk of nosocomial pneumonias caused by other microorganisms (which also occurred earlier after the onset of mechanical ventilation), thereby confirming that nosocomial pneumonia caused by *Acinetobacter* is a late complication of mechanical ventilation in a subset of very debilitated patients requiring prolonged respiratory assistance.

6.3 PREDISPOSING FACTORS

A number of factors have been suspected or identified as increasing the risk of pneumonia or colonisation of the lower respiratory tract by *Acinetobacter* in the ICU, including advanced age, chronic lung disease, immunosuppression, surgery, use of antimicrobial agents, presence of invasive devices such as endotracheal and gastric tubes, and type of

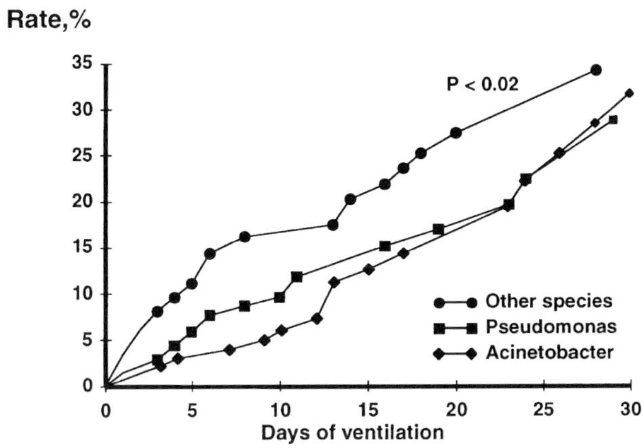

FIGURE 6.2
Cumulative rate of ventilator-associated pneumonia in 159 consecutive patients, subdivided according to causative bacterial species (data taken from Trouillet et al., 1995).

respiratory equipment (Buxton et al., 1978; Castle et al., 1978; Peacock et al., 1988; Bergogne-Bérézin and Joly-Guillou, 1991; Struelens et al., 1993; Lortholary et al., 1995). Although many of these variables are likely to be interrelated, only a few studies of potential risk factors have used appropriate statistical modelling to define independent risk determinants (Townsend, 1987).

During an 18-month period in 1983 and 1984, aminoglycoside-resistant *Acinetobacter* were isolated from 98 patients at a university hospital in North Carolina. In response to this persistent outbreak, Peacock et al. (1988) initiated an extensive epidemiological investigation to identify risk factors for the acquisition of the epidemic organism. A case-control analysis was used in which ICU patients with respiratory colonisation or infection with *Acinetobacter* were compared with matched ICU controls containing other Gram-negative bacilli in their sputum. Compared with controls, the duration of ICU stay before colonisation or infection with *Acinetobacter* was significantly longer for cases (14.7 days vs. 5.9 days; $p = 0.002$). Although exposure to invasive devices and procedures was not significantly different for the two groups, cases received respiratory therapy for significantly longer than did controls (14.7 days vs. 6.6 days; $p = 0.006$). Aminoglycoside usage in the two groups was comparable, but cases tended to have received an aminoglycoside for a longer duration before colonisation or infection than had controls (9.0 days vs. 6.1 days; $p = 0.08$) and they had also received more cephalosporins than controls (1.9 vs. 1.2; $p = 0.018$) (Table

TABLE 6.3

Case-Control Analysis of Nosocomial Colonisation or Infection with Multi-Resistant *Acinetobacter* in an ICU

	Case (n = 22)	Controls (n = 22)	*p* Value
Duration of ICU stay (days)	14.7	5.9	0.002
Duration of ventilator use (days)	14.7	6.6	0.006
No. of antibiotics received	5.4	3.4	0.012
No. of cephalosporins received	1.9	1.2	0.018
Duration of aminoglycoside use (days)	9.0	6.1	0.08

Note: Data taken from Peacock et al. (1988).

6.3). In a similar study of 40 patients infected or colonised with *A. baumannii* in which the cases were compared by logistic regression analysis with 348 non-infected and non-colonised patients who were present in the ICU at the same time, Lortholary et al. (1995) also demonstrated that the severity of the underlying disease, as evaluated by the APACHE II score (Knaus et al., 1985), and the presence of a previous infection that required antimicrobial treatment were independent factors for acquiring *A. baumannii.*

There is therefore little doubt that extended ICU care as a result of severe underlying disease, prolonged respiratory therapy with mechanical ventilation, and previous antimicrobial therapy are all key factors in predisposing to this type of infection. Since the only factor amenable to prevention in this setting is antimicrobial therapy, avoidance of unnecessary antibiotics should be a high priority in management of such patients. Other studies have confirmed that these so-called superinfections presumably occur as a consequence of selection of more resistant bacterial pathogens, such as *Acinetobacter* or *Pseudomonas* spp., during treatment with antibiotics. In a study conducted by Fagon et al. (1989b) of a large group of 567 consecutive patients receiving mechanical ventilation, it was noted that *Pseudomonas* or *Acinetobacter* spp. were responsible for 65% of the pneumonias among patients who had received broad-spectrum antimicrobial drugs, but for only 19% of pneumonias among patients who had not received antibiotics previously (Figure 6.3).

Several other studies have demonstrated a clear relationship between antibiotic use and the prevalence of multiresistant isolates. Go et al. (1994) reported a nosocomial outbreak of infections caused by imipenem-resistant *A. baumannii* that occurred in a New York hospital after increased use of imipenem for therapy of cephalosporin-resistant *Klebsiella* infections. Using restriction endonuclease analysis to further characterise the various isolates of *A. baumannii* collected during the

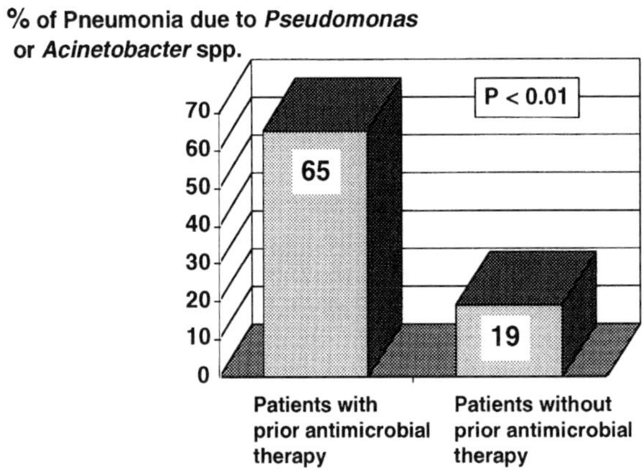

FIGURE 6.3
Relationship between previous antimicrobial therapy and the percentage of total pneumonias caused by *Acinetobacter* or *Pseudomonas* spp. (data taken from Fagon et al., 1989b).

survey, it was demonstrated that *Acinetobacter* isolates resistant to imipenem alone, or to imipenem and amikacin, were identical, but that these differed from broadly susceptible isolates, suggesting that colonisation of staff and the environment by organisms within any of these multiresistant antibiotype groups may have contributed to the imipenem-resistant *A. baumannii* outbreak. Indeed, it was entirely possible that a previous multiresistant clone of *A. baumannii* served as a precursor to the imipenem-resistant isolate, becoming progressively resistant to imipenem and amikacin during environmental transmission.

6.4 NOSOCOMIAL TRANSMISSION

As described in Chapter 4, numerous outbreaks of multiply resistant *A. baumannii* infections have been reported in surgical and medical ICUs. Nosocomial spread of *A. baumannii* in the ICU setting has been associated particularly with contaminated ventilatory equipment and with colonised nursing and respiratory therapy staff via hand transmission (see Chapter 4). However, in many outbreaks, the epidemiology of nosocomial respiratory colonisation and/or infection with *A. baumannii* is now increasingly complex because of the coexistence of epidemic cases with unrelated sporadic cases caused by different strains. This underlines the necessity to use molecular typing methods to improve the detection of micro-epidemics amenable to early control. In an investigation performed by Hartstein et al. (1990) in five ICUs of one hospital, where neither the frequency nor rate of *Acinetobacter* isolation exceeded the hospital endemic norms, only 10 of the 30 isolates that were studied by plasmid DNA fingerprinting had identical or similar fingerprints, suggesting that these isolates were acquired nosocomially and transmitted to multiple patients in multiple ICUs, while all other isolates examined were unrelated to this group of 10 strains. Similarly, in a study of 20 patients with nosocomially acquired *A. baumannii* infection, ribotyping of the isolates showed by the diversity of their banding patterns that an epidemic of 11 cases was co-existing with sporadic cases (Richet et al., 1994). Following case-control analyses, the only independent risk factor for epidemic *Acinetobacter* infection was to have undergone a surgical procedure in the emergency operating room before admission to the ICU, and the only risk factor for sporadic *Acinetobacter* infection was previous receipt of a fluoroquinolone. Without the use of sophisticated molecular techniques to identify the precise strain responsible for each infection, the epidemiology of this outbreak would have been impossible to determine, and efficacious control measures would have been difficult to implement.

6.5 MORTALITY AND MORBIDITY

Crude mortality rates of 30–75% have been reported for nosocomial pneumonia caused by *A. baumannii*, with the highest rates reported in ventilator-dependent patients (Fagon et al., 1989b; Torres et al., 1990; Bergogne-Bérézin and Joly-Guillou, 1991). It is therefore clear that the prognosis associated with this type of infection is considerably worse than that associated with other Gram-negative or Gram-positive bacteria, with the exception of *P. aeruginosa*. In the series of patients studied by Fagon et al. (1989b), in which the diagnosis was retained only if PSB specimens grew $>10^3$ cfu of at least one organism/ml, mortality associated with *Pseudomonas* or *Acinetobacter* pneumonia was >75%, compared with only 55% for pneumonias caused by other organisms ($p < 0.05$).

Although these statistics indicate that nosocomial pneumonia caused by *Acinetobacter* is a severe disease in ventilator-dependent patients, previous studies have not demonstrated clearly that pneumonia actually causes increased mortality. Two independent factors have made this important question difficult to resolve unambiguously. First, as indicated above, numerous studies have demonstrated that severe underlying illness predisposes patients to the development of pneumonia caused by *Acinetobacter*. Thus, it is difficult to establish whether such critically ill patients would have survived if nosocomial pneumonia had not occurred. Second, it is not always easy to differentiate colonised patients from patients with true pneumonia (Chastre and Fagon, 1994). Using reliable techniques, including PSB and BAL, to identify patients with ventilator-associated pneumonia, Fagon et al. (1993a) undertook a cohort study in which patients who had developed pneumonia caused by *Acinetobacter* and/or *Pseudomonas* spp. were matched carefully with control subjects for the severity of underlying illness and other important variables, such as age, indication for ventilatory support and duration of exposure to risk. The results indicated that the attributable mortality in cases of infection caused by nonfermentative organisms exceeded 40%, with a corresponding relative risk of death of 2.50 (Table 6.4).

In another study of 388 consecutive ICU patients, of whom 40 developed colonisation and/or infection with *A. baumannii* during their hospitalisation, Lortholary et al. (1995) also demonstrated that the occurrence of colonisation or infection with this type of microorganism was associated independently with death ($p < 0.01$), even when other prognostic factors such as age, presence of an ultimately or rapidly fatal underlying condition and disease severity, as rated by the APACHE II score (Knaus et al., 1985), were entered into the logistic regression model. Therefore, a large body of evidence indicates that nosocomial pneumonia caused by *Acinetobacter* is indeed associated with mortality rates in excess of those attributable to any underlying disease alone.

It is difficult to evaluate accurately the morbidity and excess costs associated with nosocomial pneumonia caused by *Acinetobacter*. However, with respect to morbidity measures, several studies have found that this type of pneumonia lengthens the ICU stay by several days (Fagon et al., 1993a; Go et al., 1994; Lortholary et al., 1995). In the study cited above, the median length of stay in the ICU for patients with ventilator-associated pneumonia caused by *Acinetobacter* or *Pseudomonas* spp. was 21 days, compared with a median for controls of 14 days (Fagon et al., 1993a). These prolonged hospitalisations underscore the considerable financial burden imposed by the development of this type of infection.

6.6 CLINICAL FEATURES

Pneumonia caused by *A. baumannii* shares many features with that caused by a number of other Gram-negative bacilli, particularly *P. aeruginosa*. Associated symptoms and signs include a combination of fever, leukocytosis and purulent tracheal secretions, with the appear-

TABLE 6.4

Crude Mortality, Attributable Mortality and Relative Risk of Death in Patients with Pneumonia Caused by *Acinetobacter* or *Pseudomonas* spp.

	Patients	%	95% C.I.
Crude mortality, cases	15/21	71.4	
Crude mortality, controls	6/21	28.6	
Attributable mortality		42.8	16.5 – 69.1
Risk ratio			2.01 – 4.57

Note: Data taken from Fagon et al. (1993a).

ance of new infiltrates on the chest roentgenogram. Descriptions of radiographic patterns have included patchy infiltrates, diffuse bronchopneumonia and nodular infiltrates (Buxton et al., 1978; Fagon et al., 1989a; Peacock et al., 1988; Trouillet et al., 1995). *Acinetobacter* pneumonia is usually multi-lobar, and frequently involves the lower lobes, but there is no evidence that nosocomial ventilator-associated pneumonia caused by this microorganism is radiologically distinct from that caused by other bacteria. Pleural effusions are common, as well as pulmonary cavitation, but the frequency with which this occurs is unknown.

As summarised in Table 6.5, which compares the clinical features observed in 105 episodes of ventilator-associated pneumonia according to bacterial origin, there were no specific hallmarks of infection caused by *Acinetobacter* that can be used in clinical practice to identify patients with this type of infection (Trouillet et al., 1995). The duration of mechanical ventilation before the onset of pneumonia was, however, longer in patients with pneumonia caused by *Acinetobacter* or *Pseudomonas* spp. compared to that for patients with pneumonia caused by other microorganisms. The radiological score and the PaO_2/FiO_2 ratio were also more severe in cases of infection caused by *Acinetobacter*, reflecting the severity of the pulmonary disease in these patients.

The most difficult task in this setting when *Acinetobacter* is isolated from tracheal aspirates is to differentiate patients who are only colonised from patients with true pneumonia. As indicated previously, most critically ill patients have serious underlying disease, increased oropharyngeal colonisation with hospital flora, including *Acinetobacter*, and numerous reasons for elevated body temperature and leukocytosis. Chest radiographic changes consistent with pneumonia may be caused by pulmonary oedema, pulmonary haemorrhage or atelectasis (Wunderink et al., 1992). Therefore, studies evaluating the usefulness of clinical parameters and/or tracheal secretions in identifying ventilated patients with nosocomial pneumonia have generally been disappointing (Fagon et al., 1988; 1993b). Accurate and prompt microbiological

TABLE 6.5

Clinical Findings in 105 Episodes of Ventilator-Associated Pneumonia According to Bacterial Origin

	Acinetobacter (n = 25)	P. aeruginosa (n = 29)	Other spp. (n = 51)
Duration of prior MV[a] (days)	28	25	16[b]
Temperature, °C	38.3	38.3	38.2
Blood leukocytes ($\times 10^3/mm^3$)	20.2	15.8	16.3
Radiologic score	6.7[b]	5.6	5.1
PaO_2/FiO_2 (mmHg)	180[b]	232	210
Total no. of alveolar cells, $\times 10^3$/ml	1076	1013	2174

[a] Mechanical ventilation.

[b] Significant difference ($p < 0.05$) compared with other groups.

diagnosis should therefore be based on reliable techniques, such as BAL and/or the PSB technique (Chastre and Fagon, 1994).

6.7 TREATMENT

Treatment of acute pneumonia caused by *A. baumannii* requires antibiotics and supportive care. Generally, two antibiotics with anti-*Acinetobacter* activity are given parenterally, but no consensus has been reached concerning issues as basic as the optimal antimicrobial regimen for therapy or the duration of treatment. Similarly, the efficacy of endotracheal or aerosolised antibiotics as either the sole or adjunctive therapy for *Acinetobacter* pneumonia remains controversial. Indeed, to date, evaluation of various antimicrobial strategies for the treatment of bacterial pneumonia in mechanically ventilated patients has been difficult for several reasons.

First, as indicated above, obtaining a definitive diagnosis of pneumonia in critically ill patients is far from easy (Chastre and Fagon, 1994). Although it is difficult to distinguish clinically between bacterial colonisation of the tracheobronchial tree and true nosocomial pneumonia, nearly all previous therapeutic investigations have relied solely on clinical diagnostic criteria, and therefore have probably included patients who did not have pneumonia. Second, most of these studies have used cultures of tracheal secretions as the major source of samples for microbiological analysis, despite the fact that the upper respiratory tract of most ventilated patients is usually colonised with multiple potential pathogens. Finally, the lack of an adequate technique to sample the infection site in the lung directly has hampered study of the ability or inability of antibiotics to eradicate the causative pathogens from the

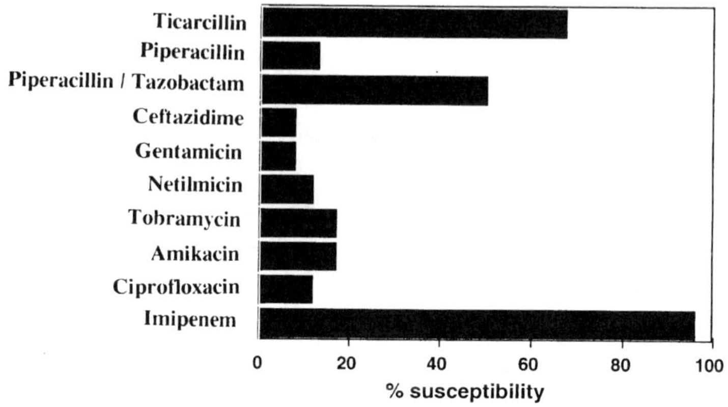

FIGURE 6.4
Antimicrobial susceptibilities of *A. baumannii* isolates recovered from patients with nosocomial pneumonia (data taken from Trouillet et al., 1995).

lower respiratory tract, and therefore to predict their efficacy. As with many other opportunistic Gram-negative bacilli, increasing antibiotic resistance has hindered the therapeutic management of nosocomial pneumonia caused by *Acinetobacter* spp. *A. baumannii* isolates are now often resistant to most available antibacterial drugs, with some centres reporting up to 80% of strains resistant to all aminoglycosides (Figure 6.4; see also Chapter 7). Even resistance to imipenem, which was for several years the most effective drug for treating *Acinetobacter* nosocomial infections, has now been described (Go et al., 1994). A full description of the different antibiotic resistance patterns of *A. baumannii* can be found in Chapter 7. Because bactericidal synergy against *Acinetobacter* has been demonstrated when carbenicillin and an aminoglycoside are combined, the use of an effective ß-lactam (e.g., piperacillin, ticarcillin, ceftazidime or imipenem) and aminoglycoside combination remains the preferred approach, whenever this is possible. Unfortunately, the unique propensity of *A. baumannii* to develop resistance to multiple antimicrobial agents reinforces concerns about the imminence of a post-antimicrobial era in which no effective antibiotics will be available to treat this type of infection.

ACKNOWLEDGMENT

The authors wish to thank C. Brun and A. Failin for their invaluable help in the preparation of this chapter.

REFERENCES

Allardet-Servent, A., Bouziges, N., Carles-Nurit, M., Bourg, G., Gouby, A. and Ramuz, M. (1989). Use of low-frequency-cleavage restriction endonucleases for DNA analysis in epidemiological investigations of nosocomial bacterial infections. *J. Clin. Microbiol.* 27, 2057–2061.

Allen, D.M. and Hartman, B.J. (1990). *Acinetobacter* species. In *Principles and Practice of Infectious Diseases*, pp. 1696–1700. G.L. Mandell, R.G. Douglas and J.E. Bennet (Eds). Churchill Livingstone, New York.

Bergogne-Bérézin, E. and Joly-Guillou, M.-L. (1991). Hospital infection with *Acinetobacter* spp.: an increasing problem. *J. Hosp. Infect.* 18 (Suppl. A), 250–255.

Bouvet, P.J.M. (1991). Typing of *Acinetobacter*. In *The Biology of* Acinetobacter, pp. 37–52. K.J. Towner, E. Bergogne-Bérézin and C.A. Fewson (Eds). Plenum Press, New York.

Buxton, A.E., Anderson, R.L., Werdegard, D. and Atlas, E. (1978). Nosocomial respiratory tract infection and colonization with *Acinetobacter calcoaceticus*: epidemiologic characteristics. *Am. J. Med.* 65, 507–513.

Castle, M., Tenney, J.H., Weinstein, M.P. and Eickhoff, T.C. (1978). Outbreak of a multiply resistant *Acinetobacter* in a surgical intensive care unit: epidemiology and control. *Heart Lung.* 7, 641–644.

Cefai, C., Richards, J., Gould, F.K. and McPeake, P. (1990). An outbreak of *Acinetobacter* respiratory tract infection resulting from incomplete disinfection of ventilatory equipment. *J. Hosp. Infect.* 15, 177–182.

Centers for Disease Control (1987). Nosocomial infection surveillance, 1984. *Surv. Summaries.* 35 (no. 155), 17SS–29SS.

Chastre, J. and Fagon, J.Y. (1994). Pneumonia in the ventilator-dependent patient. In *Principles and Practice of Mechanical Ventilation*, pp. 857–890. M.J. Tobin (Ed.). McGraw-Hill, New York.

Cook, D.J., Fitzgerald, J.M., Guyatt, G.H. and Walter, S. (1991). Evaluation of the protected brush catheter and bronchoalveolar lavage in the diagnosis of pneumonia. *J. Intensive Care Med.* 6, 196–205.

Craven, D.E., Barber, T.W., Steger, K.A. and Montecalvo, M.A. (1990). Nosocomial pneumonia in the 1990s: update of epidemiology and risk factors. *Semin. Respir. Infect.* 5, 157–172.

Cunha, B.A., Klimek, J.J., Gracewski, J., McLaughlin, J.C. and Quintiliani, R. (1980). A common source outbreak of *Acinetobacter* pulmonary infections traced to Wright respirometers. *Postgrad. Med. J.* 56, 169–172.

Fagon, J.Y., Chastre, J., Hance, A.J., Guiguet, M., Trouillet, J.L., Domart, Y., Pierre, J. and Gibert, C. (1988). Detection of nosocomial lung infection in ventilated patients. Use of a protected specimen brush and quantitative culture techniques in 147 patients. *Am. Rev. Respir. Dis.* 138, 110–116.

Fagon, J.Y., Domart, Y., Chastre, J., Trouillet, J.L., Montravers, P. and Gibert, C. (1989a). Is death of patients with ventilator-associated *Pseudomonas* pneumonia related with pulmonary infection or with patient conditions. *Proc. 5th World Congr. Intensive and Critical Care Medicine (Kyoto).*

Fagon, J.Y., Chastre, J., Domart, Y., Trouillet, J.L., Pierre, J., Darne, C. and Gibert, C. (1989b). Nosocomial pneumonia in patients receiving continuous mechanical ventilation. Prospective analysis of 52 episodes with use of a protected specimen brush and quantitative culture techniques. *Am. Rev. Respir. Dis.* 139, 877–884.

Fagon, J.Y., Chastre, J., Hance, A.J., Montravers, P., Novara, A. and Gibert, C. (1993a). Nosocomial pneumonia in ventilated patients: a cohort study evaluating attributable mortality and hospital stay. *Am. J. Med.* 94, 281–288.

Fagon, J.Y., Chastre, J., Hance, A.J., Domart, Y., Trouillet, J.L. and Gibert, C. (1993b). Evaluation of clinical judgment in the identification and treatment of nosocomial pneumonia in ventilated patients. *Chest.* 103, 547–553.

Go, E.S., Urban, C., Burns, J., Kreiswirth, B., Eisner, W., Mariano, N., Mosinka-Snipas, K. and Rahal, J.J. (1994). Clinical and molecular epidemiology of *Acinetobacter* infections sensitive only to polymyxin B and sulbactam. *Lancet* 344, 1329–1332.

Hartstein, A.I., Rashad, A.L., Liebler, J.M., Actis, L.A., Feeman, J., Rourke, J.W., Stibolt, T.B., Tomalsky, M.E., Ellis, G.R. and Crosa, J.H. (1988). Multiple intensive care outbreak of *Acinetobacter calcoaceticus* subspecies *anitratus* respiratory infection and colonization associated with contaminated reusable ventilator circuits and resuscitation bags. *Am. J. Med.* 85, 624–631.

Hartstein, A.I., Morthland, V.H., Rourke, J.W., Freeman, J., Garber, S., Sykes, R. and Rashad, A.L. (1990). Plasmid DNA fingerprinting of *Acinetobacter calcoaceticus* subspecies *anitratus* from intubated and mechanically ventilated patients. *Infect. Control Hosp. Epidemiol.* 11, 531–538.

Knaus, W.A., Draper, E.A., Wagner, D.P. and Zimmerman, J.E. (1985). APACHE-II: a severity of disease classification system. *Crit. Care Med.* 13, 818–829.

Lortholary, O., Fagon, J.Y., Buu Hoi, A., Slama, M.A., Pierre, J., Giral, P., Rosenzweig, R., Gutmann, L., Safar, M. and Acar, J. (1995). Nosocomial acquisition of multiresistant *Acinetobacter baumannii*: risk factors and prognosis. *Clin. Infect. Dis.*, 20, 790–796.

Peacock, J.E., Sorrell, L., Sottile, F.D., Price, L.E. and Rutala, W.A. (1988). Nosocomial respiratory tract colonization and infection with aminoglycoside-resistant *Acinetobacter calcoaceticus* var *anitratus*: epidemiologic characteristics and clinical significance. *Infect. Control Hosp. Epidemiol.* 9, 302–308.

Richet, H., Villers, D., Coste-Borel, M., Richard, P., Le Gallou, F., Espaze, E. and Nicolas, F. (1994). *Acinetobacter baumannii* infections in ICU patients: coexistence of epidemic and sporadic cases. *Abstr. 34th Intersci. Conf. Antimicrobial Agents and Chemotherapy (Orlando).*

Schaberg, D.S., Culver, D. and Gaynes, R. (1991). Major trends in the microbial etiology of nosocomial infections. *Am. J. Med.* 91, 72S–75S.

Stone, J.W. and Das, B.C. (1985). Investigation of an outbreak of infection with *Acinetobacter calcoaceticus* in a special care baby unit. *J. Hosp. Infect.* 6, 42–48.

Struelens, M.J., Carlier, E., Maes, N., Serruys, E., Quint, W.G.V. and van Belkum, A. (1993). Nosocomial colonization and infection with multiresistant *Acinetobacter baumannii*: outbreak delineation using DNA macrorestriction analysis and PCR-fingerprinting. *J. Hosp. Infect.* 25, 15–32.

Torres, A., Aznar, R., Gatell, J.M., Jiminez, P., Gonzalez, J., Ferrer, A., Celis, R. and Rodriguez-Roisin, R. (1990). Incidence risk and prognosis factors of nosocomial pneumonia in mechanically ventilated patients. *Am. Rev. Respir. Dis.* 142, 522–528.

Townsend, T.R. (1987). New aspects of case-control studies and clinical trials. In *Prevention and Control of Nosocomial Infections,* 1st ed., pp. 578–590. R.P. Wenzel (Ed.). Williams & Wilkins, Baltimore.

Trouillet, J.L., Vuagnat, A., Calvat, S., Joly-Guillou, M.-L., Dombret, M.C., Clavier, H., Gibert, C. and Chastre, J. (1995). Ventilator-associated pneumonia due to *Acinetobacter baumannii:* prospective analysis of 22 episodes. Abstr. American Thoracic Society. International Conference, Seattle.

Vandenbroucke-Grauls, G.M., Kerver, A.J., Rommes, J.H., Ransen, R., den Dekker, C. and Verhoef, J. (1988). Endemic *Acinetobacter anitratus* in a surgical intensive care unit: mechanical ventilators as reservoir. *Eur. J. Clin. Microbiol. Infect. Dis.* 7, 485–489.

Wunderink, R.G., Woldenberg, L.S., Zeiss, J., Day, C.M., Ciemins, J. and Lacher, D.A. (1992). The radiologic diagnosis of autopsy-proven ventilator-associated pneumonia. *Chest* 101, 458–463.

Chapter 7

Resistance of *Acinetobacter* spp. to Antimicrobials — Overview of Clinical Resistance Patterns and Therapeutic Problems

E. Bergogne-Bérézin

CONTENTS

0-8493-9223-3/95/$0.00+$.50

7.1 INTRODUCTION

In most developed countries, about 5–10% of all hospitalised patients develop a nosocomial infection responsible for increased morbidity and mortality, prolonged stays in intensive care units (ICUs), and a subsequent increased economic burden for hospitals. Noticeable changes in the pattern of hospital-acquired infection over the past two decades have resulted from various factors, including the selective pressure exerted by broad-spectrum antimicrobial agents and environmental factors (Gould, 1994). Thus the organisms responsible for nosocomial infections are evolving continuously, although the predominant sites of nosocomial infection remain more or less unchanged, with nosocomial pneumonia, urinary tract infection, post-operative wound infection and bacteraemia accounting for about 80% of reported infections. In this context, among emerging Gram-negative aerobic non-fermentative bacilli, *Acinetobacter* spp. have become important pathogens (Bergogne-Bérézin and Joly-Guillou, 1985), with several recent investigations demonstrating that they are now responsible for 5% of all nosocomial infections, and that they account for 9–10% of all pathogens involved in nosocomial pneumonia (Joly-Guillou et al., 1992b; Struelens et al., 1993). A dramatic increase in the number of publications concerning *Acinetobacter* spp. (microbiology, epidemiology, resistance to antibiotics, nosocomial infections, etc.) in the medical and scientific literature has occurred since the first few articles in the 1960s, so that >1600 articles had been published by 1990. This increase reflects the emerging interest of clinicians, epidemiologists and microbiologists in this particular genus. A particular concern has been the frequent multiple antibiotic resistance exhibited by nosocomial acinetobacters, and the resulting therapeutic problems involved in treating patients with nosocomial infections in ICUs.

This chapter not only deals with clinically significant resistance to antibiotics exhibited by *Acinetobacter* spp., as described either in outbreaks of infection or as a result of *in vitro* studies, but also examines particular therapeutic problems and possible therapeutic solutions available from the current antibacterial armamentarium. Throughout this chapter, the taxonomic nomenclature used by the original authors cited has been retained. Readers are referred to Appendix I at the end of this book for details of the early classification schemes used for *Acinetobacter* spp., and their probable correlation with the latest taxonomic designations.

7.2 HISTORICAL ASPECTS

7.2.1 Background to Resistance Development

Several major phenomena may account for the continually evolving clinical problem of bacterial resistance to antibiotics: (1) occurrence of mutations in genes located either on the bacterial chromosome or on plasmids; (2) transfer of resistance genes into 'new' bacterial hosts that were previously susceptible; (3) emergence of 'new' acquired resistance mechanisms; and (4) selection of 'new' opportunistic pathogens that are multiresistant to antibiotics (Courvalin, 1994). The latter phenomenon, if applied to *Acinetobacter* spp., suggests that these bacteria were originally inherently resistant to antibiotics in nature, and that they have been selected in hospitals by intensive use of antibiotics. The extent to which the other three phenomena account for the evolution of resistance in *Acinetobacter* spp. is discussed in more detail in Chapter 8. However, several major questions can be addressed by examining the history of emerging antibiotic resistance in *Acinetobacter* spp. In particular: Were the first recognised clinical isolates resistant to available antibiotics in the 1950s–1960s? What was the resistance phenotype of *Acinetobacter* strains isolated from the non-clinical environment? Are the *Acinetobacter* strains involved in community-acquired infections similar to those isolated in hospitals with regard to their susceptibility/resistance patterns? Is there a correlation between the use of modern broad-spectrum antibiotics and the resistance observed in *Acinetobacter* spp.? This chapter attempts to answer these questions, based on published data and a lengthy personal experience of dealing with these organisms.

7.2.2 Evolution of Resistance in *Acinetobacter* spp.

All recent publications on *Acinetobacter* infections state repeatedly that "*Acinetobacter* spp. are saprophytic organisms of increasing pathogenic importance" and that "*Acinetobacter* has developed one of the most impressive patterns of antibiotic resistance during the last 20 years" (French et al., 1980; Larson, 1984; Bergogne-Bérézin and Joly-Guillou, 1985; Buisson et al., 1990; Joly-Guillou et al., 1992a; Struelens et al., 1993), and that "an increasing number of reports in the medical literature document the high rates of resistance in *Acinetobacter*." Many studies of the precise mechanisms of resistance to ß-lactams, aminoglycosides, fluoroquinolones and the other major antibiotic classes are in progress and, with the proliferation of molecular biology methods and modern technology, many mechanisms and genetic determinants of resistance have been elucidated within the past few years (see Chapter 8).

The numerous published studies devoted to resistance problems in *Acinetobacter* spp., extending over more than 20 years from the initial observations of *Acinetobacter* infections to the current situation, can be divided schematically into two main periods: (1) an early period of discovery and increasing awareness of the role of *Acinetobacter* spp., and their susceptibility to the antibiotics available between 1960 and 1974 (see Section 7.3); and (2) a secondary period during which ICUs became particularly important, with an increasing use of invasive procedures (catheters, central lines, exploratory techniques, etc.) and powerful new antibiotics, including new aminoglycosides, imipenem, third-generation cephalosporins and fluoroquinolones. This second period has lasted from 1975 until today, during which time new patterns of resistance have emerged and been disseminated in many hospitals (see Section 7.4).

7.3 INITIAL LEVELS OF RESISTANCE IN *Acinetobacter* spp.

7.3.1 Resistance in Strains Isolated from the Environment

Most studies and reviews deal with clinical resistance in *Acinetobacter* spp., i.e., resistance to antibiotics in strains isolated either from patients with a nosocomial infection or from the hospital environment, with the latter being the reservoir or source of outbreak strains with multiresistance patterns (see Chapter 4). The question arises as to whether *Acinetobacter* strains isolated from soil or water are inherently resistant, and whether they possess preexisting resistance genes mediating resistance to most antibiotic classes.

It is well known that Gram-negative bacilli possess a tough outer membrane layer outside the cell-wall peptidoglycan that limits antibiotic access into the cell, and that these bacteria are less permeable to large antibiotic molecules than Gram-positive bacteria. As a result, many Gram-negative species are inherently resistant to some antibacterial agents. Thus *Acinetobacter* has been described as "intrinsically insensitive" to ß-lactams, especially aminopenicillins and cephalosporins (Goldstein et al., 1983), but this observation was made by studying an epidemic strain isolated in a hospital. Does this statement apply to strains isolated far from any medical and human contact? Does it apply to acinetobacters isolated from community-acquired infections? In the 50 years since antibiotics were introduced for therapeutic use, many bacterial species, including *Acinetobacter* spp., have acquired a variety of resistance mechanisms. In the pre-antibiotic era, the opportunity to express resistance did not exist in clinical settings, but the preexistence of resistance genes in the environment is an accepted fact, although experimental data regarding antibiotic resistance in free-living saprophytic organisms in nature are sparse in the scientific literature. Examination of soil samples from a remote site in the Solomon Islands revealed a few bacteria containing R-plasmids (Gardner et al., 1969). Faecal carriage of antibiotic-resistant bacteria in normal individuals who have never been treated with antibiotics, and the occurrence of antibiotic-resistant strains in animals, may result either from the 'natural' presence of resistance genes in bacteria in soil and water, or from non-clinical use of antibiotics (e.g., antibiotic feed supplements given to animals) (Linton, 1986). Naturally occurring plasmids have been described in *Acinetobacter* (Towner, 1991), and genetic exchanges observed between soil and aquatic bacteria are also likely to occur with *Acinetobacter* spp. in natural ecosystems (Courvalin, 1994). Thus it can be postulated that various potential mechanisms of resistance in *Acinetobacter* have existed for many years, but that these have only become expressed in clinical conditions as a result of the selective pressures exerted by consecutively introduced antibiotics in therapy.

7.3.2 Early Susceptibility Data for Clinical Isolates

Looking back at the very first data on the susceptibility/resistance of *Acinetobacter* clinical isolates, obtained at a time when very few antibiotics were available and patients were treated with combinations of penicillin and either streptomycin, tetracycline, erythromycin or chloramphenicol, the outcome was not always favourable (Table 7.1) and death occurred in more than half of the published case reports (Courtieu and Chassignol, 1961). The strains isolated in these and other studies were tested for their susceptibilities to antibiotics by means of non-standardised techniques at a time when disc diffusion tests, 'tablets' and dilution techniques were 'hand-made' (Table 7.2). In these studies published in the 1950s and early 1960s, at a time when the selective pressure exerted today by broad-spectrum antibiotics did not exist in hospitals, it is noticeable that for all strains tested (without clear delineation of species), a general observation was made of resistance to penicillin and chloramphenicol, whereas tetracyclines (oxytetracycline and chlortetracycline) were generally active against most strains, while streptomycin, erythromycin and sulphonamides showed variable activity (Table 7.2); all strains were inhibited by polymyxin, colistin and neomycin (Chabbert and Courtieu, 1961; Courtieu and Chassignol, 1961). As shown in Table 7.3, two 'new' macrolides, spiramycin and oleandomycin, introduced for therapeutic use in the late 1950s, were more active against *A. lwoffii* than against *A. glucidolytica*. It had already been observed that there was a difference in susceptibility to antibiotics between the then recognised varieties: i.e., *A. lwoffii* was more susceptible than *A. glucidolytica*. Some antibiotic combinations studied *in vitro* (i.e., chloramphenicol + erythromycin; streptomycin + tetracycline or erythromycin), were either synergic or indifferent, while others (e.g., chloramphenicol + tetracycline) were antagonistic. Several conclusions can be drawn from these historical data:

1. Despite taxonomic uncertainties, the pathogenic role of *Acinetobacter* spp. in hospitals was already starting to be recognised;
2. Despite a limited range of available antimicrobial agents with a predominantly bacteriostatic mode of action (i.e., tetracyclines and macrolides), opportunistic pathogens had already emerged, although the antibiotics in use were probably unable to modify the local microbial ecology significantly;

TABLE 7.1

Early Antibiotic Therapy in Cases of *Acinetobacter* Infection (Courtieu and Chassignol, 1961)

Year	Origins[a]	Age (years)	Organisms[b+]	Antibiotic therapy	Outcome
1949	Pneumonia	52	*Mima polymorpha*	Sulphonamide + penicillin	Death
1951	Bacteraemia	8	*Mima polymorpha*	Tetracycline	Cure
1956	Pneumonia	35	*Moraxella lwoffii glucidolytica*	Penicillin + streptomycin	Death
				chloramphenicol + erythromycin	Cure
1956	Endocarditis ?	19	*Herella vaginicola*	Penicillin + streptomycin	Death
	Bacteraemia			Erythromycin + streptomycin + tetracycline + sulphonamide	
1957	Bacteraemia	62	*B. anitratum*	Tetracycline + streptomycin + erythromycin + chloramphenicol	Death
	Bacteraemia	4	*B. anitratum*	Penicillin + streptomycin chloramphenicol	Cure
1958	Pneumonia	Premature	*B. anitratum*	Pencillin + streptomycin	Death
	Pneumonia	Premature	*B. anitratum*	Penicillin + streptomycin	Death
	Pneumonia	Premature	*B. anitratum*	Aureomycin + erythromycin + chloramphenicol	Death
	Pneumonia	Premature	*B. anitratum*	Penicillin + streptomycin	Death
1958	Bacteraemia	33	*Mima polymorpha*	Penicillin + tetracycline	Cure
1959	Pneumonia	50	*B. anitratum*	Penicillin + streptomycin tetracycline + chloramphenicol	Death

[a] Underlying disease not stated (predominantly surgical conditions).

[b] Isolated from blood cultures.

TABLE 7.2

Susceptibility Data in Early Studies

Year	**1951**		**1953**		**1954**		**1955**		**1959**	
No. of strains	**86**		**24**		**77**		**20**		**20**	
(%)	**S**	**R**	**S**	**R**	**S**	**R**	**S**	**R**	**S**	**R**
Penicillin	—	85 (99%)	—	R	—	100%	—	R	—	R
Streptomycin	49	37 (43%)	—	R	61%	39%	—	—	S	—
Chloramphenicol	—	85 (99%)	—	R	—	95%	—	R	S	—
Sulphathiazole	V*	—	—	—	40%	—	—	—	S	—
Tetracycline	—	—	—	—	—	—	—	—	S	—
Oxytetracycline	76 (88%)	—	S	—	98%	—	S	—	—	—
Chlortetracycline	81 (88%)		S	—	100%	—	S	—	S	—
Erythromycin	—	—	40%	—	12%	—	—	—	S	—
Spiramycin	—	—	—	—	—	—	—	—	—	R
Polymyxin	—	—	—	—	—	—	S	—	S	—
Neomycin	—	—	—	—	100%	—	—	—	S	—
Novobiocin	—	—	—	—	—	—	—	—	S	—
Identification	*B5 W*		*Herellea*		*Bacterium anitratum*		*Diplococcus mucosus*		*M. glucidolytica*	

*Variable.

Data from Courtieu, A.L. and Chassignol, S. (1961). *Les Germes du Groupe* Moraxella (Acinetobacter), Bacterium anitratum. *Fréquence. Bactériologie*. Fayolle, Lyon.

TABLE 7.3

Comparison of Susceptibility/Resistance of *Acinetobacter (Moraxella) lwoffii* and *Acinetobacter glucidolytica*

	Acinetobacter (Moraxella) lwoffii				*Acinetobacter (Moraxella) glucidolytica*			
No. strains	93		10		83		28	
	Non-proteolytic		**Proteolytic**		**Non-proteolytic**		**Proteolytic**	
	% S	% R*	% S	% R*	% S	% R*	% S	% R*
Penicillin	6.4	93.5	—	100	—	100	—	96.4
Streptomycin	48.3	51.6	60	40	48.2	51.8	60.7	39.2
Chloramphenicol	97.8	—	100	—	30.1	69.8	85.7	14.2
Tetracycline	97.8	—	100	—	73.1	26.8	96.4	—
Oxytetracycline	97.8	—	100	—	69.8	30.1	85.7	14.2
Chlortetracycline	100	—	100	—	81.9	18.0	96.4	—
Erythromycin	100	—	90	—	91.5	8.4	89.2	10.7
Spiramycin	97.8	—	70	—	71	28.9	78.5	21.4
Polymyxin	100	—	100	—	100	—	96.4	—
Neomycin	100	—	100	—	98.7	—	96.4	—
Novobiocin	—	61	—	100	24	76	—	—
p-Amino-phenylsulphonamide	94.6	—	90	—	60.2	39.7	89.2	10.7
Sulphonamides								
Sulphathiazole	94.6	—	90	—	56.6	43.3	89.2	10.7
Sulphamethylthiodiazole	94.6	—	90	—	56.6	43.3	89.2	10.7
Sulphadiazine	93.5	—	90	—	59.0	40.9	89.2	10.7

* Resistant strains include those of intermediate susceptibility.

Data from Courtieu, A.L. and Chassignol, S. (1961). *Les Germes du Groupe* Moraxella (Acinetobacter), Bacterium anitratum. *Fréquence. Bactériologie.* Fayolle, Lyon.

3. Amongst these pathogens, *A. glucidolytica* was described in several studies as showing variable susceptibility to available antibiotics in the 1950s to 1960s, and was resistant by definition to penicillin G (Courtieu and Chassignol, 1961; Table 7.3);
4. Bacteria designated 'B5W', *Bacterium anitratum*, *Moraxella glucidolytica* (non-proteolytic) were susceptible to erythromycin, novobiocin and tetracyclines, but were predominantly resistant to chloramphenicol, generally resistant to penicillin, and were stated to be penicillinase producers (Chabbert and Courtieu, 1961; Courtieu and Chassignol, 1961; Tables 7.2 and 7.3).

The differences in susceptibilities between species or varieties shown in Table 7.3 had already been noted, so that the so-called variety *M. glucidolytica* was considered to be more resistant to tetracyclines, macrolides, sulphonamides and chloramphenicol than *M. lwoffii*. It had also been recognised that proteolytic strains of *M. glucidolytica* (now likely to correspond to *A. haemolyticus*) showed a major difference in that most strains (85–100%) were susceptible to chloramphenicol, streptomycin, tetracyclines, erythromycin, spiramycin, polymyxin and neomycin.

Thus, by this time, the role of *Acinetobacter* spp. as nosocomial pathogens had begun to be recognised, albeit in somewhat sparse and occasional reports. Treatments were often disappointing (Table 7.1), and the susceptibility/resistance of pathogenic strains had already become a major concern. In another 10–20 years, the development of broad-spectrum antibiotics was to lead to an exponential increase in the number of major problems caused by *Acinetobacter* infections in hospitals (see Section 7.4).

7.3.3 Resistance in Strains Isolated from Community-Acquired Infections

Rare cases of community-acquired pneumonia, urinary tract infection, or native valve endocarditis have been described in the medical literature (see Chapters 5 and 6).

Pneumonia — Case reports and available susceptibility data are summarised in Table 7.4. In most cases of pneumonia, various risk factors, particularly alcoholism (9 of 14 cases in one series), were identified (Goodhart et al., 1977; Rudin et al., 1979; Cordes et al., 1981); these are discussed in more detail in Chapter 6. In these early reports, as shown in Table 7.4, *Acinetobacter* isolates were susceptible to tetracyclines, aminoglycosides and carbenicillin; they were resistant to chloramphenicol, ampicillin, penicillin and first-generation cephalosporins (e.g., cephalothin), but were susceptible to cefuroxime and to

TABLE 7.4

Susceptibility and Antibiotic Therapy in Community-Acquired Pneumonia Caused by *Acinetobacter* spp.

References	No.	Age	Case reports: risk factors	Antibiotic therapy	Susceptibility/resistance MIC (mg/L)	Outcome
Goodhart et al. (1977)	1	33	Alcoholic smoker	(1) Cephalothin i.v. 8 g (2) Cephalothin 12 g + gentamicin 5 mg/kg	R: Chloramphenicol Penicillins Cephalosporins	Died
	2	58	Alcoholic	(1) Penicillin G i.v. 2.4 Miu (2) Ampicillin i.v. + gentamicin	S: Minocycline (0.09) Gentamicin (3.1) Tobramycin (0.78) Amikacin (0.78) R: Chloramphenicol (50) Tetracycline (6.25)	Cured
Rudin et al. (1979)	3	58	Alcoholic Asthmatic	Clindamycin Gentamicin	—	Died
	4	41	Alcoholic Acute pancreatitis	Penicillin Gentamicin Carbenicillin	—	Survived

TABLE 7.4 (continued)

Susceptibility and Antibiotic Therapy in Community-Acquired Pneumonia Caused by *Acinetobacter* spp.

References	No.	Age	Case reports: risk factors	Antibiotic therapy	Susceptibility/resistance MIC (mg/L)	Outcome
Rudin et al. (1979)	5	35	Alcoholic	Penicillin Gentamicin	—	Died
	6	51	Lymphoma Chemotherapy	Gentamicin Carbenicillin	—	Survived
	7	74	None	Penicillin Gentamicin	—	Survived
	8	44	Alcoholic Chronic bronchitis	(1) Penicillin – gentamicin (2) Gentamicin – carbenicillin	—	Survived
1959[a]	9	50	None	(1) Penicillin, streptomycin, sulfonamide (2) Tetracycline, chloramphenicol	S: Aminoglycosides Tetracycline	Survived
1968	10	49	None	Penicillin	Carbenicillin	Died
1968	11	49	Alcoholic	Penicillin	R:	Died
1973	12	69	Chronic renal failure	(1) Cephalothin – gentamicin (2) tetracycline – gentamicin – carbenicillin	Chloramphenicol Cephalosporins Penicillin	Survived
1976	13	50	Alcoholic	gentamicin – carbenicillin	Ampicillin	Survived
1977	14	58	Alcoholic	(1) Penicillin – gentamicin (2) Carbenicillin	Methicillin	Survived

Cordes et al. (1981)		3 cases	Foundry workers	Recommended therapy: Carbenicillin + gentamicin Cefuroxime, cefotaxime[b]		2 Died
				Antibiotic therapy		
				Initial	**Secondary**	
Anstey et al. (1992)[c]	1,5, 6,8,9	5 cases	Alcoholics/ smokers	Ampicillin	—	4 Died
	2	1 case	Alcoholic	Kanamycin Chloramphenicol Penicillin	TMP-SMZ Gentamicin	Died
	3	1 case	Alcoholic, smoker, diabetic	Penicillin	TMP-SMZ Gentamicin	Died
	4	1 case	Heavy smoker	Ampicillin Gentamicin	Cefotaxime Gentamicin	Cured
	7	1 case	Smoker, alcoholic	Cefotaxime Piperacillin	—	Cured
	10	1 case	Smoker, alcoholic	Ceftriaxone	Gentamicin Penicillin Ceftazidime	Died
	11	1 case	Smoker, alcoholic diabetic	Penicillin Gentamicin	Ciprofloxacin Imipenem	Cured

Note: TMP-SMZ — trimethoprim-sulfamethoxazole; Miu — million international units.

[a] Reported by Rudin et al. (1979).

[b] Appelbaum, P.C. et al. (1981). Sensitivity of *Acinetobacter calcoaceticus* strains to seven cephalosporins. *Lancet* ii, 96.

[c] Seven strains tested in this series: sensitive to: piperacillin, imipenem, gentamicin, carbenicillin; resistant to: cefotaxime.

cefotaxime, the first available third-generation cephalosporin (Cordes et al., 1981). Besides risk factors, the use of inappropriate antibiotics (e.g., cases 10 and 11 of Rudin's series in Table 7.4) was responsible for therapeutic failure. Until 1980, the limited number of available compounds contributed to the prescription of inappropriate (ineffective) antibiotics. A more recent report (Anstey et al., 1992) of 11 cases of community-acquired pneumonia and 34 compiled cases from the literature underlined again the high mortality rate (64%) associated with this disease and the limited choice of antibiotics used in combinations until 1977. Extended spectrum penicillins (e.g., piperacillin), third-generation cephalosporins, imipenem and aminoglycosides were the drugs recommended for use in combinations by Anstey et al. (1992). Failure occurred whenever an unsuitable antibiotic, such as penicillin or ampicillin, was used.

Endocarditis — In 15 cases of native valve endocarditis, 40% of patients had no underlying pathology (Table 7.5) and, surprisingly, early cases (No. 1 in 1951; No. 2 in 1956; Table 7.5) were cured by oxytetracycline or combinations including erythromycin, tetracycline and streptomycin. Of the total cases, 66% were cured with various antibiotic combinations, but the latest reported isolate, identified as *A. anitratus,* was susceptible only to gentamicin, cefotaxime and imipenem, with the latter antibiotic being used successfully in monotherapy (Gradon et al., 1992). Although acinetobacters are usually reported to be resistant to chloramphenicol, one case (No. 13), was treated successfully with chloramphenicol. The precise species identification was not available, and non-*A. baumannii* strains may have been included in these reports; such strains would have been susceptible to antibiotics that were inactive against *A. baumannii* (e.g., chloramphenicol).

Urinary tract infection — In a large study of community-acquired urinary tract infection (Hoffmann et al., 1982), a 14% incidence of patients with significant bacteriuria caused by *A. calcoaceticus* was reported; susceptibility testing indicated that the *A. calcoaceticus* strains were insensitive to ampicillin, cephalosporins, chloramphenicol, erythromycin and streptomycin. Carbenicillin, oxytetracycline, sulphonamides and tetracyclines showed variable activity. The compiled susceptibility data since 1964 (Table 7.6) do not indicate significant changes in resistance to antibiotics actually available during the years of observation until 1980. Biovar *anitratus* strains were more resistant than biovar *lwoffii,* as shown for strains tested in 1975 and in 1982.

It should be emphasised that the basic rates of resistance were very low for all these series of strains isolated in the community, except for penicillin G, cephalothin and chloramphenicol, so that most antibiotics, including ampicillin, all aminoglycosides, co-trimoxazole, rifampicin

TABLE 7.5

Susceptibilities and Antibiotic Therapy in Native Valve Endocarditis Caused by *Acinetobacter* spp. (Based on the Data of Gradon *et al.*, 1992)

Case reports				Antibiotic		
No.	Age	Underlying pathology	Valve affected	therapy (susceptibilities)	Outcome	Year of reference
1	8	CHD	—	Oxytetracycline	Cured	1951
2	19	RHD	A	Erythromycin Tetracycline Streptomycin	Cured	1956
3	23	RHD	A	Chloramphenicol Penicillin	Died	1963
4	52	Bicuspid	A	None	Died	1964
5	31	None	A	Chloramphenicol Ampicillin Tetracycline	Cured	1966
6	18	RHD	M	Penicillin	Cured	1967
7	32	None	A	Ampicillin Chloramphenicol	Died	1971
8	13	None	A	Penicillin Gentamicin (Isoniazid)	Died	1987
9	46	None	?	Gentamicin Cephapirin	Cured	1979
10	11	CHD	?	—	Cured	1980
11	22	CHD	M	—	Cured	1980
12	26	RHD	—	Chloramphenicol	Cured	1980
13	27	None	M + A	Imipenem	Died	1981
14	48	None	M	Imipenem[a]	Cured	1992

CHD, congenital heart disease; RHD, rheumatic heart disease; A, aortic; M, mitral.

[a] Strain of *A. anitratus* susceptible to gentamicin, imipenem and cefotaxime (Gradon et al., 1992).

and nalidixic acid, were still active in 1982. These compiled data were published in 1982 when some newer broad-spectrum antibiotics (cefotaxime, ceftazidime, amikacin, imipenem, ticarcillin, piperacillin, etc.) were available, but were being used only in hospitals. This suggests that most strains present in the community were susceptible to antibiotics used in general practice (even to ampicillin and cephalothin), as well as to compounds prescribed only in hospitals.

In a later epidemiological study (Gerner-Smidt, 1987), 328 *Acinetobacter* isolates were collected from clinical specimens in hospitals and from general practice. Of the hospital strains, 99% were resistant to ampicillin vs. 46% from general practice; 75% were resistant to carbenicillin vs. 42% from general practice; and 57, 19 and 18% of hospital strains were resistant to gentamicin, sulphonamides and

TABLE 7.6

A. calcoaceticus Antibiotic Susceptibilities

Year	No. of strains	% of strains[a] susceptible to AM	CA	CE	PE	ST	AK	NE	TO	GE	KA	CL	ER	DO	MI	OY	TE	LI	RI	CO	SU	SX	NA
1964	68	0	—	—	0	0	—	—	—	—	35	0	0	—	—	—	0	—	—	56	—	—	—
1965	19 – 54[b, c]	—	—	—	0	17	—	—	—	—	93	6	4	—	—	4	2	—	—	80	20	—	—
1975	23 – 142[b]	9	49	1	—	44	—	—	—	90	95	18	22	—	—	—	49	—	—	84	—	—	96
1975	249[d]	—	—	—	0	77	—	—	—	—	83	54	—	—	—	—	68	—	—	89	—	—	—
	75[e]	—	—	—	0	58	—	—	—	—	81	75	—	—	—	—	83	—	—	78	—	—	—
1976	65[f]	32	—	2	—	—	—	—	—	98	—	—	—	—	95	—	26	—	—	100	—	—	—
1977	100	—	—	—	—	—	—	—	—	70	95	0	—	—	—	—	97	—	—	78	56	—	—
1979	694	3	—	0	—	—	55	—	77	54	83	2	—	97	—	—	38	—	—	94	—	—	—
1980	305 – 1,208[b, c]	13	64	6	—	—	—	—	88	87	88	17	—	—	—	—	66	—	—	96	90	—	—
1982	107[c]	96	75	49	2	91	88	99	99	100	96	83	98	100	100	98	96	3	100	99	94	96	100
1982	14[d]	86	43	21	0	93	93	100	100	100	100	21	86	100	100	93	93	0	100	100	100	100	100
	93[e]	98	80	53	2	90	87	99	99	100	96	53	100	100	100	99	97	3	100	99	94	96	100

Note: AM, ampicillin; CA, carbenicillin; CE, cephalothin; PE, penicillin; ST, streptomycin; AK, amikacin; NE, netilmicin; TO, tobramycin; GE, gentamicin; KA, kanamycin; CL, chloramphenicol; ER, erythromycin; DO, doxycycline; MI, minocycline; OY, oxytetracycline; TE, tetracycline; LI, lincomycin; RI, rifampicin; CO, colistin; SU, sulphonamides; SX, sulphamethoxazole-trimethoprim; NA, nalidixic acid.

[a] If not stated otherwise, all strains were of biovar *anitratus*.

[b] The agents were tested against various numbers of strains, the ranges of which are indicated.

[c] Both biovars were represented.

[d] *A. anitratus*.

[e] *A. lwoffii*.

[f] 64 Biovar *anitratus* strains and one biovar *lwoffii* strain were tested.

Based on data from Hoffmann, S., Mabeck, C.E. and Vejsgaard, R. (1982). *J. Clin. Microbiol.*, 16, 443–451.

tetracyclines, respectively, vs. none from general practice. In the same study, 90% of *A. anitratus* hospital strains were multiresistant to more than five antibiotics vs. 50% from general practice that were susceptible to all antibiotics tested and only 21% that were resistant to five antibiotics. The activity of antibiotics towards *Acinetobacter* strains involved in hospital infections is described in more detail below.

7.3.4 Early Clinical Resistance in Hospitals

In the early 1970s, *Acinetobacter* infections at the Bichat Hospital in Paris were treated successfully with gentamicin, minocycline, nalidixic acid, ampicillin or carbenicillin, either as single agents or in antibiotic combinations. In early *in vitro* studies, the majority of *Acinetobacter* clinical isolates were susceptible to ampicillin (60–70%), gentamicin (92.5%), chloramphenicol (57%) and nalidixic acid (97.8%) (Bergogne-Bérézin et al., 1971). Of 86 *Acinetobacter* strains isolated in 1969, only 3.4% were resistant to carbenicillin. In the same study, a comparison between strains isolated from medical wards (120 isolates) and those from surgical wards or ICUs (120 isolates) showed a significant difference in susceptibilities, with the former showing higher proportions of isolates susceptible to ampicillin (80 vs. 41.6%), cephaloridine (65.8 vs. 29.1%) and streptomycin (84.1 vs. 42.5%). The difference in gentamicin susceptibility between the two groups of strains was smaller; indeed the vast majority of *Acinetobacter* isolates were susceptible to gentamicin (95.8% of strains from medical wards vs. 87.5% of those from ICUs).

Strains of *A. calcoaceticus* var. *anitratus* were significantly more resistant than var. *lwoffii* strains in early studies (Bergogne-Bérézin et al., 1971; Pinter and Kantor, 1971), a finding confirmed by more recent investigations. In a further study with 64 strains isolated in 1973–1974, increased rates of resistance were noted compared with 1971 (Table 7.7), and it is interesting to note that soon after the therapeutic introduction of a new powerful antibacterial agent in the form of carbenicillin, the frequency of resistant strains had risen from 3.5 to 7.2%. Simultaneously, resistance to ampicillin had risen from 39 to 43.3%, and to cephaloridine from 52.5 to 66.6%. There was no change in resistance rates for the other available antibiotics, streptomycin, kanamycin and chloramphenicol, between 1971 and 1974, probably because of the reduced use of these older drugs as newer, more active compounds became available. With tetracyclines, *A. glucidolytica* was less resistant in 1971 to minocycline (67.3%) than to tetracycline (98.5%). These early studies at Bichat Hospital provided data that were similar to the previous studies cited above, and were followed in 1975 by the start of regular surveillance of *Acinetobacter* strains isolated from our own and other hospitals in France (see next section).

TABLE 7.7

Evolution of Resistance Between 1971 and 1973–74

	% Resistant strains		
Antibiotics	**1971 (n=240)**	**1974 (n=64)**	
Ampicillin	39	43.3	
Cephaloridine	52.5	66.6	
Carbenicillin	3.5	7.2	
Streptomycin	36.7	28.3	
Kanamycin	20.5	23.3	*A.glucidolytica*
Gentamicin	7.5	8.3	
Chloramphenicol	43.0	43.0	
Colistin	3.4	1.6	
Tetracycline	98.5	87.5	
Minocycline	67.3	3.1	
Tetracycline	—	72.7	*A. lwoffii*
Minocycline	—	45.5	1972 (n = 11)

Based on E. Bergogne-Bérézin, unpublished data.

7.4 EVOLUTION OF RESISTANCE AND OVERALL CURRENT RESISTANCE PATTERNS IN *Acinetobacter* spp.

Resistance to antimicrobial agents is a matter of degree. Thus, a favourable outcome in a patient may be obtained even although careful analysis of a given strain may indicate that it carries a resistance gene. For instance, a gene encoding ß-lactamase production may be present, but may not be expressed at a level sufficient to cause treatment failures. This chapter is concerned with the *clinical resistance of Acinetobacter*, based on an analysis of susceptibility/resistance patterns. The precise *mechanisms* and *genetic basis* of the antibiotic resistance observed in *Acinetobacter* spp. are discussed in Chapter 8.

7.4.1 Overview of Current Resistance Patterns

Since 1975, successive surveys have shown increasing resistance in clinical isolates of *Acinetobacter* spp., and this has been reported periodically (Garcia et al., 1983; Joly-Guillou and Bergogne-Bérézin, 1985; Obana et al., 1985; Godineau-Gauthey et al., 1988). High proportions of strains have become resistant to older antibiotics; indeed, many *Acinetobacter* strains are now resistant to most commonly used antibacterial drugs, including aminopenicillins, ureidopenicillins, cephalosporins of the first (cephalothin) and second (cefamandole) generations (Morohoshi and Saito, 1977; Joly-Guillou and Bergogne-Bérézin, 1985), cephamycins such as cefoxitin (Garcia et al., 1983), most aminoglycosides-aminocyclitols (Dowding, 1979; Devaud et al., 1982; Goldstein et al., 1983; Joly-Guillou and Bergogne-Bérézin, 1985), chloramphenicol and tetracyclines. Contrasting data are sparsely distributed in the literature for some antibiotics: minocycline and doxycycline were cited in one study as the most active of 21 antibiotics tested, including 10 ß-lactams, 3 aminoglycosides and nalidixic acid (Obana et al., 1985), while piperacillin and aminoglycosides were effective in a study of the *in vitro* activity of 25 antibacterial agents (Garcia et al., 1983). These differences in susceptibilities of *Acinetobacter* isolates could be attributable to environmental factors such as the different patterns of antibiotic usage that exist in Japan (Obana et al., 1985), the U.S. (Garcia et al., 1983) or France (Joly-Guillou and Bergogne-Bérézin, 1985).

Among various factors associated with the increasing resistance of *Acinetobacter* spp. to antibiotics, an increased incidence of the more resistant var. *anitratus* (more or less corresponding to *A. baumannii*) may also have changed the reported overall antibiotic resistance of clinical isolates of *Acinetobacter*. Other *Acinetobacter* spp., i.e., *A. lwoffii*, *A. johnsonii*, *A. haemolyticus* and *A. junii*, can be isolated from the hospital environment, but are involved less frequently in nosocomial infection and show variable susceptibility to antibacterial agents (Bergogne-Bérézin et al., 1987; Freney et al., 1989). Most studies have focused on the resistance exhibited by the *A. calcoaceticus* var. *anitratus* (*A. baumannii*) strains that are responsible for most outbreaks of nosocomial infection associated with this genus. The emergence of antibiotic resistance and current resistance patterns in this species are summarised in Table 7.8.

7.4.2 Clinical Resistance in Outbreaks

Many outbreaks of *Acinetobacter* nosocomial infection and endemic episodes have been described in the medical literature, and the resistance patterns of the responsible strains have been reported, based generally on standard antibiotic susceptibility testing by the disc diffusion method. Table 7.9 summarises the results of some of these studies, scanning a period of 1979–1993.

Multiresistance in *Acinetobacter* spp. — Initially, this involved resistance to ampicillin, first-generation cephalosporins and chloramphenicol, with strains remaining susceptible to carbenicillin, gentamicin, tobramycin and tetracyclines. However, since 1980, resistance to carbenicillin, ureidopenicillins (mezlocillin and azlocillin) and newer cephalosporins (cefuroxime, cefsulodin and cefotaxime) has been observed in *Acinetobacter* isolates (French et al., 1980; Devaud et al., 1982; Joly-Guillou and Bergogne-Bérézin, 1985).

TABLE 7.8

Evolution of Resistance and Current Percentages of Resistant Strains (No. of Strains Tested) of *A. baumannii*

	Bichat-Claude Bernard University Hospital[a]							France[a]	Germany[b]	Spain[c]
	1975	1980	1985	1991	1992		1993[d]	1991	1993	1993
	(260)	(67)	(218)	(354)	(302)		(217)	(NS[e])	(95)	(54)
	%	%	%	%	%[f]	% ICU	%	%	%	%
ß-Lactams										
Ampicillin	83	94	98	100	—	—	43	—	91	98
Carbenicillin/ticarcillin	16	70	72	77.5	40.4	46	45	44	—	70
Mezlocillin/piperacillin	—	80	92	88	82.7	94	NS	—	36	67
Aztreonam	85	93	90	90	NS	NS	6.9	—	54	98
Imipenem	—	—	0	2	2.3	5	NS	0.6	0	0
2nd-Gener. cephalosporins	90	95	98	98	NS	NS	95.4	—	99	—
Cefotaxime	—	55	84.5	95	91.5	98	97.3	—	32	69
Ceftazidime	—	19	45.5	92	82.4	91	94.7	76	32	45
Moxalactam	—	90	90	92	NS	NS	—	—	—	—
Aminoglycosides										
Gentamicin	12	81	84	80	74	88	78	69.5	98	67
Tobramycin	12	22	47	76	58	61	65	62	98	50
Netilmicin	—	16	49	85	64.5	79	73	44.5	—	34
Amikacin	—	2.5	39	78	68	81	69	55	64	28
Kanamycin	17	67	70	89	—	—	—	—	—	—
Fluoroquinolones										
Pefloxacin			19	75	75	83	73	63	—	—
Ciprofloxacin			18	70	—	—	79	—	94	30
Ofloxacin			20	67	—	—	77.5	—	—	28
Norfloxacin			87	95	—	—	—	—	—	82

Note: NS, not stated.

[a] Data from Joly-Guillou et al., 1992a. [b] Data from Seifert et al., 1993. [c] Data from Vila *et al.*, 1993. [d] Unpublished results. [e] Compiled data from 32 diagnostic laboratories.

TABLE 7.9

Clinical Resistance in Outbreak Strains or in Strains Isolated from Endemic Cases of *Acinetobacter* Infections

No. of strains[a] (duration of outbreak)	Pathology or origin (colonisation)	Susceptibility/resistance pattern	Ref.
1372 *A. calcoaceticus* (1974–77)	% Endemic 29 Respiratory tract 27 Urines 21 Wounds 9.3 Bacteraemia	Am Ce Chl – R Car Co Ge Ka To Tet – S	Retaillau et al. (1979)
15 *A. anitratus* (16 months)	Endemic : 15 episodes of septicaemia	Am Ce Chl – R Ca Ge Ka To – R (15 – 25% R)	Ramphal and Kluge, 1979
1 *A. anitratus* (epidemic strain)	Outbreaks : 40 patients Urines Skin Environment	Am Ce Cfox Cfs Car Mez Azl – R Sm Ka Ge Tmp-Smz Tet – R Cfu Ctx – R Nal Co To Amk – S	French et al., (1980)
1 *A. anitratus* (epidemic strain)		Am Ce Ge Ka Chl Tet Sul – R To Amk – S	Devaud et al., 1982
62 *A. calcoaceticus* (2 months)	Endemic 6 Pus 35 Urines 4 Faeces 11 Respiratory tract Other : environment	Am Mez Ce Ctx Cfs Ge Nal Chl – R To Amk Tet Rif Tmp-Smz – S	Santos-Ferreira et al., 1984
328 *A. calcoaceticus* (8 months)[b] 117 *A. anitratus* 200 *A. lwoffii*	% Endemic 8.5 URT[c] 17 Respiratory tract 18.2 Urines 20 Skin 14 Female genital tract	*A. anitratus:* Am Chl Ery Car Sm – R Tet Su Ge – S (57% R) *A. lwoffii:* Am Chl Ery Car Smz – S Tet Su Ge – S	Gerner-Smidt, 1987

TABLE 7.9 (continued)

Clinical Resistance in Outbreak Strains or in Strains Isolated from Endemic Cases of *Acinetobacter* Infections

No. of strains[a] (duration of outbreak	Pathology or origin (colonisation)	Susceptibility/resistance pattern	Ref.
43 *A. calcoaceticus* var. *anitratus*	Endemic: 33 nosocomial infections 14 UTI 9 Wound infections 5 Pneumonia 5 Septicaemia	Tic Mez Ctx Caz – (83%) R Ge To Amk Tet – (42%) R (To 25% – S)	Muller-Serieys et al., 1989
228 *A. baumannii*	Endemic: 148 patients 46 Bacteraemia 16 Pneumonia 37 Wounds 30 Urines	Am Tic Ce Ctx Caz Ge Ka Tet Tmp-Smz – R Oflo To Amk Net – S	Amor et al., 1993
144 *A. calcoaceticus* var. *anitratus*	Outbreaks 90% Pneumonia 10% Bacteraemia	1989 Tmp-Smz Pol Dox Caz Imp – S 1992 Caz Imp Amk – R Pol Sulb only S	Urban et al., 1993
54 *A. baumannii*	Endemic cases	Am Az Ctx Ge Pip Tic – R Caz Cip Dox Imp Net – S Tic+Cla Tob (50%)	Vila et al., 1993
237 *Acinetobacter* spp. (6 months)	Outbreaks (285 episodes) 18 Bacteraemia 59 Respiratory tract 53 Skin/wounds 64 Urines 43 Others	Am Ce Cfu Cx Az Tm AAc Pip Ge To Amk Tmp-Smz – R (Variable multiresistant phenotypes)	Struelens et al., 1993

103 (out of 400) *A. baumannii* (6 months)	9 Outbreaks 275 patients	Ge To Am Cip Imp – S To Imp – S Amk Imp – S Imp only – S	Seifert et al., 1993
31 *A. baumannii* + 10 strains from environment (Imp - R strains)	Outbreaks (%): 31 patients 35.4 Pneumonia 22.5 Bacteraemia 12.9 Abdomen 9.6 Urines	29% Pol Ka To Amk Net – S (only) 71% Pol – S (only) (Imp – R strains)	Tankovic et al., 1994
72 *A. baumannii* (1 year) + 181 environmental strains and 121 strains from staff	Endemic (%): 59 patients 54 Pneumonia 12 Wounds 4 UTI	72% Strains: Imp + all antibiotics – R (59 patients) 1.9% Environ. strains: polymyxins – S	Go et al., 1994

Note: R = >50% Resistant; S = >70% Susceptible. AAc, amoxycillin+clavulanic acid; Am, ampicillin; Amk, amikacin; Az, aztreonam; Azl, azlocillin; Car, carbenicillin; Caz, ceftazidime; Ce, 1st-gen. cephs. (cephalothin, cephaloridin, cephalexin); Cfox, cefoxitin; Cfs, cefsulodin; Cfu, cefuroxime; Chl, chloramphenicol; Cip, ciprofloxacin; Co, colistin; Ctx, cefotaxime; Ery, erythromycin; Ge, gentamicin; Imp, imipenem; Ka, kanamycin; Mez, mezlocillin; Nal, nalidixic acid; Pip, piperacillin; Pol, polymyxin B; Sm, streptomycin; Tet, tetracycline; Tic+cla, ticarcillin+clavulanic acid; Tic, ticarcillin; Tm, temocillin; Tmp-Smz, trimethoprim-sulphamethoxazole; To, tobramycin.

[a] Nomenclature of isolates as reported in original reference.

[b] Including 56.4% strains isolated from general practice.

[c] URT = upper respiratory tract.

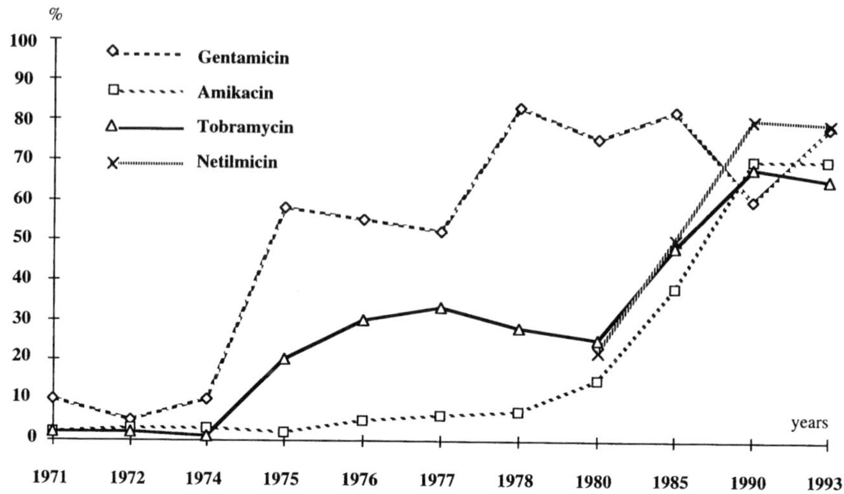

FIGURE 7.1
Resistance of *Acinetobacter* spp. to aminoglycosides 1971–1993. (Data taken from E. Bergogne-Bérézin and M.-L. Joly-Guillou, 1991; unpublished results.)

Aminoglycosides — The emergence of resistance in *Acinetobacter* spp. to aminoglycosides was followed by the rapid identification of aminoglycoside-inactivating enzymes (Bergogne-Bérézin et al., 1980; Gomez-Lus et al., 1980; Murray and Moellering, 1980). Resistance to gentamicin, streptomycin, kanamycin and tobramycin was the first to emerge, with 15–29% of strains found to be resistant in an early study by Ramphal and Kluge (1979). In Paris, a sudden rise in strains resistant to gentamicin and tobramycin occurred in 1975–1976 (Figure 7.1). The data in Table 7.9 indicate that *Acinetobacter* strains have been mostly susceptible to amikacin, but a new resistance phenotype including amikacin resistance was described by Joly-Guillou et al. (1987) and analysed in terms of its enzymic and genetic mechanism by Lambert et al. (1990).

ß-Lactams — As shown in Table 7.9, *Acinetobacter* strains have now become resistant to most commonly used ß-lactam antibiotics, including aminopenicillins, ureidopenicillins, cephalosporins of the first (cephalothin) and second (cefamandole) generations (Morohoshi and Saito, 1977; Bergogne-Bérézin and Joly-Guillou, 1991) and cephamycins (cefoxitin) (Joly-Guillou et al., 1992a), while MICs of some newer antibiotics, such as third-generation cephalosporins (cefotaxime, ceftazidime) and imipenem, have increased between 1980 and 1992. Figure 7.2 compares data from seven countries showing increasing resistance to carbenicillin/ticarcillin and ceftazidime over the period 1975–1991.

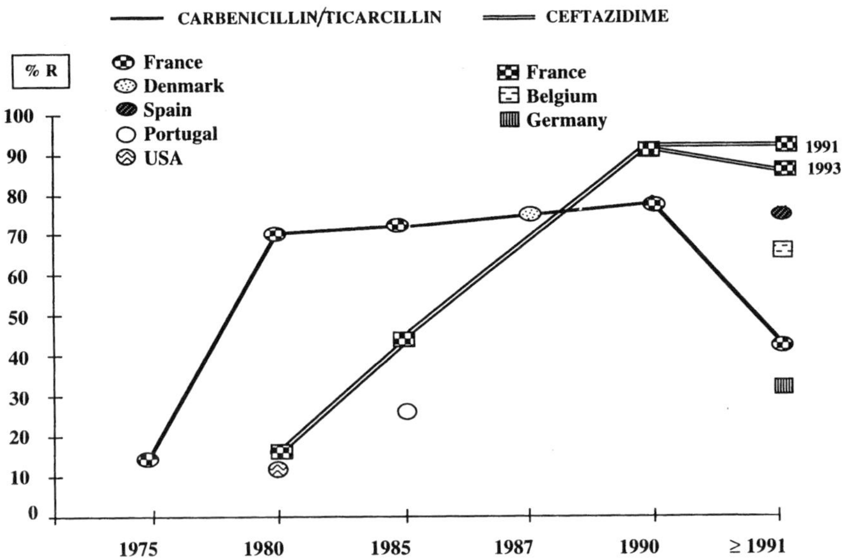

FIGURE 7.2
Evolution of resistance in *Acinetobacter* spp. to ß-lactams over an 18-year period.

Imipenem — As shown in Table 7.10, occasional imipenem-resistant strains with MICs of 4 or 8 mg/L have been isolated from patients, although one strain had an MIC of >128 mg/L. However, until recently, imipenem retained activity against 100% of strains (e.g., Muller-Serieys et al., 1989; Amor et al., 1993; Seifert et al., 1993; Vila et al., 1993); indeed, in some reports the only active drugs were imipenem and the polymyxins. Unfortunately, the most recent extensive analyses of hospital outbreaks have documented the spread of imipenem-resistant strains (Go et al., 1994; Tankovic et al., 1994). This is a particularly worrying development which threatens the continued successful treatment of *Acinetobacter* infections. Most resistance to imipenem has been observed in *A. baumannii*, while the MICs of carbapenems to non-*A. baumannii* strains has remained below 0.3 mg/L (Figure 7.3). The possible mechanisms of resistance to imipenem are discussed in Chapter 8.

Other antibiotics — In addition to resistance to ß-lactams, aminoglycosides and chloramphenicol, *Acinetobacter* strains have also shown a rapid development of resistance to fluoroquinolones. In 1985, pefloxacin, ciprofloxacin and ofloxacin inhibited about 80% of strains (Table 7.8). The frequency of selection of quinolone-resistant mutants is higher in non-fermenting organisms than in enterobacteria (see Chapter 8), and within a few years the incidence of resistant *Acinetobacter*

TABLE 7.10

Susceptibility of *Acinetobacter* spp. to Imipenem

Years	Range (mg/L)	MIC_{50}	MIC_{90}	Reference
1981	0.016 – 0.25	0.12	0.25	Phillips et al., 1981
1983	0.05 – 12.5	0.20	0.39	Garcia et al., 1983
1986	0.016 – 4.0	0.28	0.61	Bergogne-Bérézin and Joly-Guillou, 1986
1988	0.125 – >128	0.37	0.60	Joly-Guillou et al., 1990
1988	0.5 – 4.0	2.0	2.0	Chow et al., 1988
1990	0.06 – 8.0	0.25	1.75	Joly-Guillou et al., 1992b
1993	0.03 – 1.0	0.1	0.4	E. Bergogne-Bérézin, unpublished data

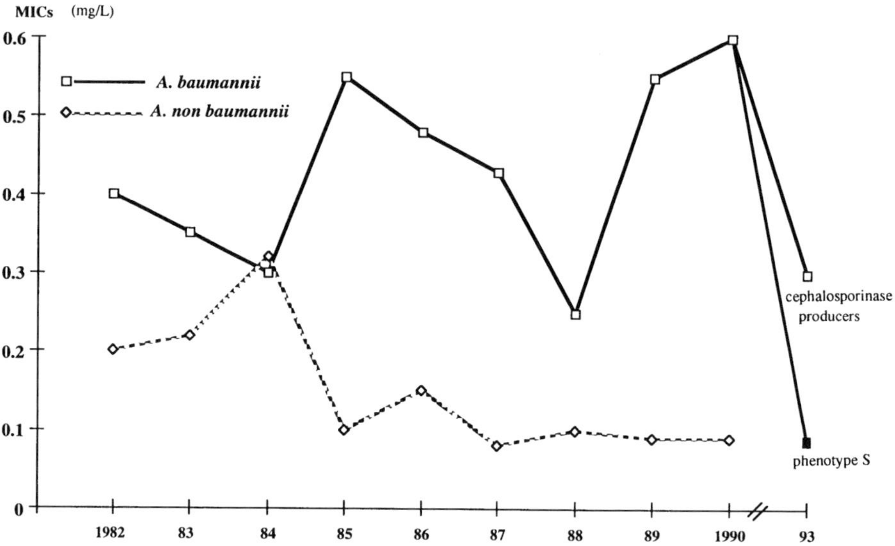

FIGURE 7.3
Susceptibility of *Acinetobacter* spp. to imipenem. (Data taken from Joly-Guillou et al., 1992b; E. Bergogne-Bérézin, unpublished results.)

strains rose from 18–20% to 75–80% in France (Joly-Guillou et al., 1992b) and to 94% in Germany (Seifert et al., 1993) (Table 7.8; Figure 7.4).

***A. baumannii* vs. non-*A. baumannii* isolates** — Species other than *A. baumannii* isolated from the hospital environment, i.e., *A. lwoffii*, *A. johnsonii* and *A. junii*, are involved less frequently in nosocomial infection and are generally less resistant to antibiotics (Gerner-Smidt, 1987; Traub and Spohr, 1989; Joly-Guillou et al., 1992a). Strains of

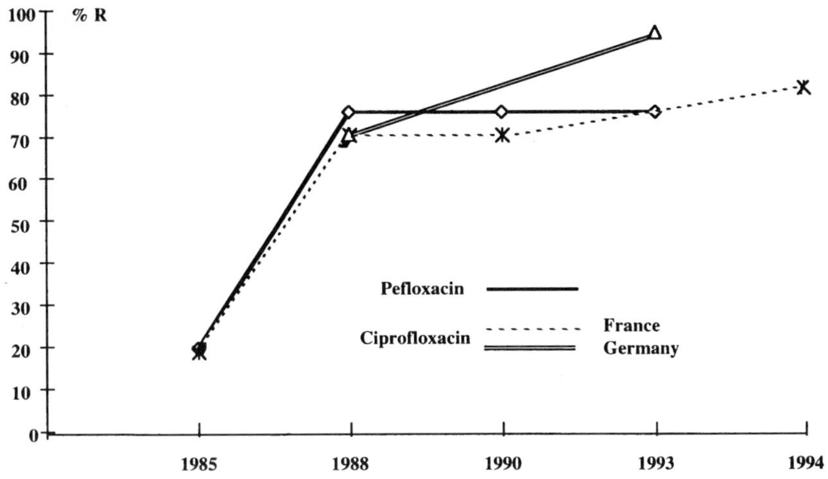

FIGURE 7.4
Resistance of *Acinetobacter* spp. to fluoroquinolones. (Data taken from Joly-Guillou et al., 1992b; Seifert et al., 1993.)

A. lwoffii are more susceptible to ß-lactams than *A. baumannii*. *A. haemolyticus* isolates are normally resistant to aminoglycosides and rifampicin, but rifampicin has mean MICs for *A. baumannii* of 2–4 mg/L and has been used effectively in synergic combination with imipenem in ICUs in France (Bergogne-Bérézin and Joly-Guillou, 1985).

7.4.3 Factors Affecting Clinical Resistance

Geographical Origin

Resistance patterns of *Acinetobacter* strains isolated from the most severe manifestations of *Acinetobacter* nosocomial infection, i.e., pneumonia, septicaemia and wound infections, have been compared to those observed in general practice or in environmental strains (Hoffmann, 1982; Gerner-Smidt, 1987). The data reported in these studies (performed some years ago) showed the expected differences between the susceptibilities of strains from hospitals and those from outside hospitals. Susceptibility patterns of acinetobacters may also vary between wards and specialities within hospitals.

Several reviews (e.g., Bousser et al., 1988; Vandenbroucke-Grauls et al., 1988) have focused on therapeutic problems in ICUs caused by *Acinetobacter* strains resistant to aminoglycosides and most ß-lactams, except for imipenem, but in one report (Urban et al., 1993), two series of isolates obtained over a period of 2–3 years had totally different susceptibility patterns. Thus, in 1989, 144 strains isolated from nosocomial infections were susceptible to co-trimoxazole, polymyxin B, doxycycline, ceftazidime and imipenem, and the authors mentioned that ceftazidime and imipenem had been prescribed widely for treating patients with *Acinetobacter* infections; in 1991–1992, strains resistant to imipenem, ceftazidime, amikacin and all other antibiotics tested routinely were isolated. Polymyxin B and sulbactam were the only agents which retained activity, so that in five patients who received ampicillin plus sulbactam, resolution of infections and eradication of the resistant organism was achieved. A high level of resistance to ampicillin and cefoperazone was observed *in vitro*, whereas these drugs combined with sulbactam provided favourable results. Similar *in vitro* activity was not observed with clavulanic acid and tazobactam against "*A. calcoaceticus* biotype *anitratus*" (Urban et al., 1993).

Data collected at Bichat Hospital in Paris are summarised in Table 7.11. The percentage of resistant strains of *A. baumannii* differed slightly between hospital wards in general and three ICUs, particularly for ceftazidime (80.7 vs. 94.7%), gentamicin (62.5 vs. 88.5%), tobramycin (51.9 vs. 76.1%) and amikacin (43.2 vs. 69.9%). These differences may be attributable to different antibiotic usage and selection pressures exerted by broad-spectrum antibiotics (e.g., aminoglycosides, imipenem and third-generation cephalosporins) that are prescribed widely in ICUs. However, it should be noted that not all resistance rates differed significantly between ICU and non-ICU strains; imipenem resistance was virtually identical (7 vs. 6.7%). Epidemiological investigations (antibiotyping, biotyping and pulsed-field gel electrophoresis) demonstrated that the *Acinetobacter* isolates collected in different ICU and non-ICU wards resulted from the dissemination of two epidemic strains with similar resistance patterns (Joly-Guillou et al., 1994a). In non-ICU wards, a reduced antibiotic selection pressure may have contributed to a lower expression of resistance to aminoglycosides and cephalosporins, but never with reversion to the original 'basic susceptibility level' of *A. baumannii*.

Antimicrobial Usage

Some studies report associations between changes in the prevalence of resistant organisms and the use of antimicrobial agents (e.g., Urban et al., 1993). It is often assumed that antibiotic usage contributes directly to the increased prevalence of resistance, but the cause-effect relationship is difficult to establish, with prospective studies needing to be designed to evaluate precisely the consumption of a given antibiotic in conjunction with surveillance of resistance rates in significant organisms during a given period of time. Our own studies with *Acinetobacter* have established such a correlation on two occasions (Joly-Guillou and

TABLE 7.11

Acinetobacters Isolated from ICUs in 1993.

		% Resistant to								
Type of Unit	**No. of strains**	**Tic**	**Ctx**	**Caz**	**Imp**	**Ge**	**To**	**Net**	**Amk**	**Pef**
Total	302 [1992]	40.4	91.5	82.4	2.3	74	58	64	68	75
Total hospital	217 [1993]	42.9	95.4	88	6.9	77.9	65	73	59	73.3
Total ICU	113[b]	45	97.3	94.7	7	88.5	76.1	78.8	69.9	78.8
Medical	34	35.3	97	91	5.9[a]	85.3	76.5	82.3	70.6	76.5
Surgical	43	55.8	100	95.4	11.6[a]	100	76.7	100	68.4	86
Infectious	36	38.9	100	97.3	2.8[a]	88.9	77.8	61	72	77.8
Total non-ICU wards	104	41.3	91.3	80.7	6.7	62.5	51.9	61.5	43.2	65.3

Note: Amk, amikacin; Caz, ceftazidime; Ctx, cefotaxime; Ge, gentamicin; Imp, imipenem; Net, netilmicin; Pef, pefloxacin; Tic, ticarcillin; To, tobramycin.

[a] Spread of imipenem-resistant strains from ICUs.

[b] Representing 52% of total isolates in 1993.

Based on E. Bergogne-Bérézin, unpublished data.

Bergogne-Bérézin, 1985; Joly-Guillou et al., 1990). Thus, antibiotic consumption was correlated with increasing resistance of *Acinetobacter* strains to nalidixic acid and amikacin for the period 1980–1985, with correlation coefficients of 0.86 and 0.79, respectively. Similarly, after several years of a large consumption of third-generation cephalosporins (cefotaxime and ceftazidime) in ICUs, the proportion of resistant *Acinetobacter* strains in 1990 had reached 97.3 and 94.7% for cefotaxime and ceftazidime, respectively (Tables 7.8 and 7.11).

Other studies have included a 7-year survey of 7385 clinical isolates in a French hospital which showed changes in the species isolated, and their antibiotic susceptibilities, as a function of consumption of third-generation cephalosporins, which increased from 1 kg to 14 kg between 1980 and 1986 (Courcol et al., 1989). The number of *A. calcoaceticus* isolates rose from 113 strains in 1980 to 250 strains in 1986, while susceptibility data indicated decreased susceptibilities to many antibiotics (Table 7.12). There was a significant relationship between the consumption of third-generation cephalosporins and the increasing number of *A. calcoaceticus* isolates, but a negative correlation between antibiotic use and resistance to aminoglycosides. It was suggested that other factors may also have been involved, such as an increase in the number of seriously ill patients and progressive changes in the bacterial ecology, resulting in the selection of strains producing aminoglycoside-modifying enzymes. However, a separate study of nosocomial outbreaks of amikacin-resistant strains of *Acinetobacter* spp. showed a clear correlation between amikacin resistance in a hospital and amikacin usage (Buisson et al., 1990).

At Bichat Hospital in Paris, the consumption of imipenem has increased in ICUs because of the prevalence of strains that are increasingly resistant to third-generation cephalosporins and other antibiotics. The subsequent spread of imipenem-resistant isolates has resulted in a significant percentage of resistant *A. baumannii* isolates in the overall survey of resistance patterns of acinetobacters in 1992–1993 (Tables 7.10 and 7.11), thereby exemplifying the statement that "changes in antimicrobial usage lead to parallel changes in the prevalence of resistance" (Finland, 1979; McGowan, 1983).

Few studies have demonstrated that decreased usage of antimicrobial agents results in a decreased prevalence of resistant organisms, but this has been observed at least once with carbenicillin/ticarcillin-resistant strains of *Acinetobacter* (Table 7.8), with lower rates of resistance in 1992–1993 (about 45%), compared with 1985–1990 (up to 77.5%). The decreased use of carboxypenicillins in response to the high prevalence of resistance to ß-lactams in ICUs may have been responsible for the subsequent reduction in resistance. As a consequence, ticarcillin can now be used again in our hospital for therapeutic purposes. Figure 7.5

TABLE 7.12

Changes in Susceptibility of *A. calcoaceticus* in Relation to Antibiotic Use

Antibiotic	Year	% Susceptible strains	Year	% Susceptible strains	Significant changes in susceptibility (*p*)	Correlation with antibiotic use (*p*)
Ticarcillin	1980	35.8	1986	9.8	<0.001	
Gentamicin		28.9		14.2	<0.001	<0.001
Tobramycin		78.8		35.1	<0.001	<0.001
Amikacin		89.7		29.9	<0.001	<0.001
Tetracycline		60.0		37.8	<0.005	
Co-trimoxazole		31.4		15.9	<0.002	

Data from Courcol, R.J., Rinkas, M. and Martin, G.R. (1989). *J. Antimicrobiol. Chemother.*, 23, 441–451.

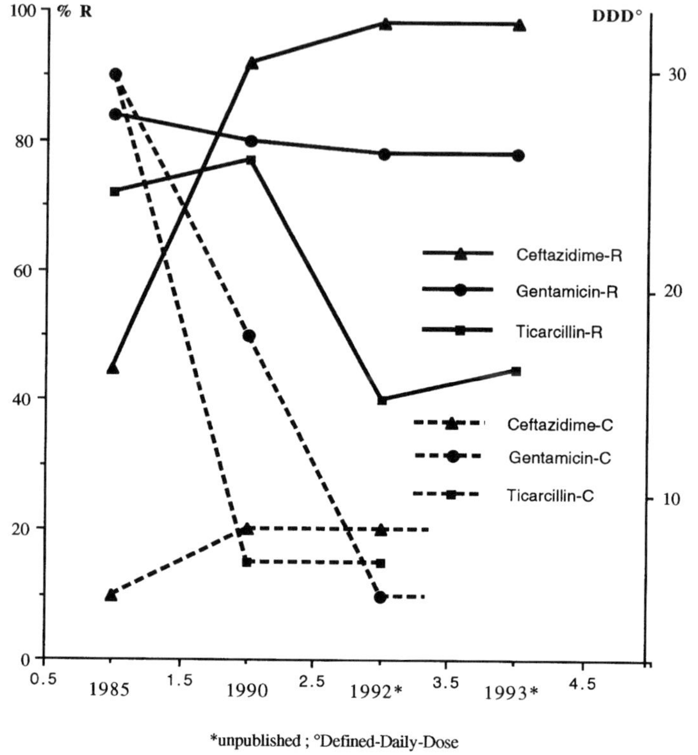

FIGURE 7.5
Antibiotic susceptibility/resistance (R) of *Acinetobacter* spp. in relation to antibiotic consumption (C). Data on antibiotic use were obtained from the records of the hospital pharmacy until 1988; consumption is expressed as 'defined daily dose' (Joly-Guillou et al., 1985). Antibiotic susceptibility data were taken from routine susceptibility testing results (see Table 7.8).

presents data collected in our hospital for the relationship between antibiotic consumption and the resistance of *Acinetobacter* strains to ticarcillin, ceftazidime and gentamicin. A significant decrease in ticarcillin-resistant strains parallels a decrease in drug consumption (with a time lag of 2–3 years); with gentamicin, even though its consumption in our hospital has been dramatically reduced, the level of gentamicin resistance in *Acinetobacter* strains has remained high, albeit with a slight decrease from 88% in 1992 to 78% in 1993.

The mechanisms by which development of resistance occurs in relation to antibiotic usage, either by means of selection of preexisting resistant clones in a bacterial population or by transfer of resistance between bacteria, or both, are discussed in Chapter 8.

7.5 RECENT *IN VITRO* SUSCEPTIBILITY/RESISTANCE DATA FOR *Acinetobacter* spp.

All susceptibility data reported in the previous sections were based on the routine susceptibility testing records of clinical laboratories. Most such data were obtained by standard disc diffusion or microdilution methods, with the results being interpreted in terms of 'susceptible' or 'resistant' strains (the latter including 'intermediate' strains) on the basis of 'breakpoints' established in different centres. Other important information on the susceptibility/resistance of *Acinetobacter* strains has been obtained from specific protocols for susceptibility testing, designed either for surveillance of clinical isolates or for the evaluation of new antimicrobial molecules in the course of development.

7.5.1 Protocol for ß-Lactam Susceptibility Surveillance

A multicentre study that included six centres in 1988, 12 centres in 1989–1990, and 13 centres in 1991–1992 was carried out in France; 3450 isolates were tested for their susceptibility/resistance to ticarcillin, cefotaxime, ceftazidime and imipenem (Figure 7.6). Interpretation of data was on the basis of breakpoints recommended for use in France (Acar et al., 1993). Over the 4-year period of surveillance, the following observations were made: (1) a slow and progressive increase in resistance of *A. baumannii* to ceftazidime, with a high starting point of 72% resistant strains; (2) a similar profile with cefotaxime, but at a higher level with up to 92% resistant strains; (3) a significant decrease in ticarcillin resistance from 53% in 1988 to 25% in 1992; and (4) a low level of resistance to imipenem, except for 1988 when 9.5% of strains were resistant or 'intermediate' to imipenem, although only 2% of strains were 'fully resistant' with an MIC of >8 mg/L. The latter observation corresponded with the emergence and spread in a few hospitals of *Acinetobacter* strains with diminished susceptibility to imipenem. The frequency of imipenem-resistant strains in 1992 was 4% (including 2.5% 'intermediate' strains) which is close to the frequency of 5% (Table 7.8) found in Bichat Hospital, Paris. Similarly, the data collected at a national level were very close to our local findings (Tables 7.8 and 7.11), particularly with regard to the returning susceptibility to ticarcillin and the high rates of resistance to third-generation cephalosporins. These findings may reflect the fact that clinical isolates of *Acinetobacter* originate predominantly from ICUs in large teaching hospitals.

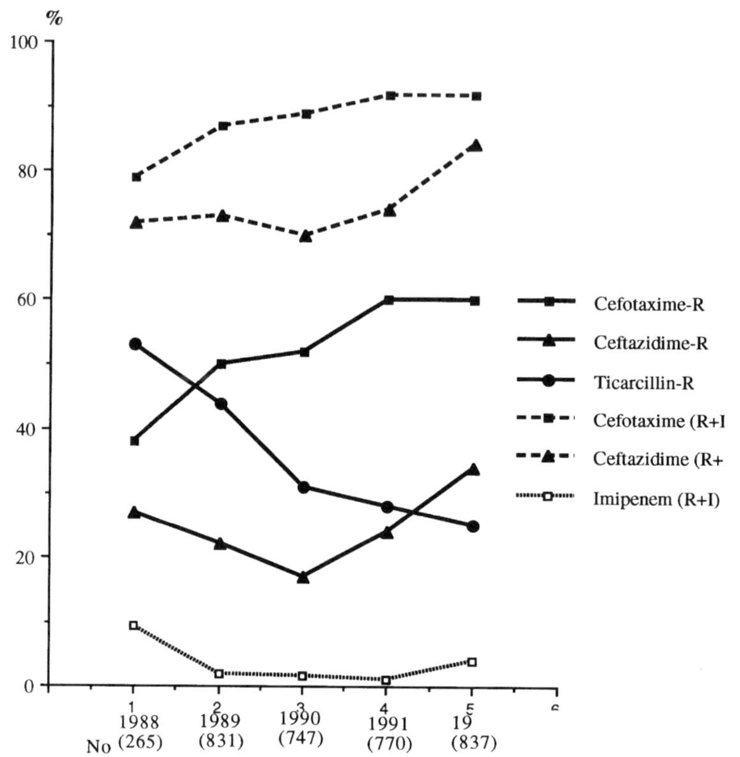

Ticarcillin-R: fully resistant with MIC >64 mg/L.
Cefotaxime & Ceftazidime-R: fully resistant with MIC >32 mg/L.

FIGURE 7.6
Resistance of *A. baumannii* (3450 isolates) to ß-lactams during 5 years of national surveillance in France.

7.5.2 Phenotypic Analysis of ß-Lactam Resistance

Joly-Guillou et al. (1988) determined the susceptibility of *A. calcoaceticus* to 14 ß-lactams and developed a phenotypic method of analysis, based on the identification of ß-lactamases involved in resistance profiles, which provided a convenient tool for further clinical studies and ß-lactam resistance surveillance. Of the 14 ß-lactams tested, resistance to combinations of ticarcillin, piperacillin, cefotaxime and ceftazidime were used to divide strains into four 'phenotypes'. A fifth phenotype was added subsequently to take account of imipenem resistance. The data in Table 7.13 summarise the consecutive phenotype figures from 9 years of surveillance, and include data from a national

TABLE 7.13

Resistance to β-Lactams and Phenotype Analysis

Phenotypes	β-lactams	Enzymes	1984–86 (n = 100) %	1988 (NS) %	1992 (n = 302) %	1993[a] (n = 217) %	1991[b] (NS) %
I	$Tic^SPip^SCtx^SCaz^S$		5	4	7.7	4.6	22
II[c]	$Tic^RPip^RCtx^SCaz^S$	Penicillinase	10	1.0	1.0	0	4
III	$Tic^SPip^VCtx^RCaz^V$	Cephalosporinase	19	19	49.6	45.5	35
IV	$Tic^RPip^RCtx^RCaz^V$	Penicillinase + Cephalosporinase	65.5	70	39.4	43	38
V + (IV)	+ $Imipenem^R$ [d]		0.5	3	2.3	6/9	1
		Total Tic^S	24	23	57.3	50.1	57
		Total cephalosporinase	84.5	89	89	88.5	73

Note: NS = not stated; Tic, ticarcillin; Pip, piperacillin; Ctx, cefotaxime; Caz, ceftazidime; ^{S} Susceptible; ^{R} Resistant; ^{V} Variable.

[a] Bichat-Claude Bernard Hospital (unpublished data).

[b] National survey.

[c] TEM-1, TEM-2, CARB-5.

[d] phenotype V (imipenem-resistant strains) may have three phenotypic profiles: I+Imp^R, III+Imp^R or IV+Imp^R.

Adapted from Joly-Guillou et al., (1988). *J. Antimicrobiol. Chemother.*, 22, 597–604 and Joly-Guillou et al., (1992b). *Abstr. 2nd Int. Conf. Prevention of Infection (Nice).*

survey carried out in 1991 (Joly-Guillou et al., 1992b). Of note are (1) the stable low proportion of totally susceptible strains (phenotype I), except for the national survey which provided a relatively high rate of susceptible isolates (22%); (2) the high proportions of cephalosporinase producers (88.5% in 1993) comprising phenotypes III and IV; and (3) the relatively high proportions of ticarcillin-susceptible strains since 1992 (57.3 and 50.1% in 1992 and 1993, respectively). These latter strains belonged to phenotypes I and III, with phenotype III being characterised by a low level of resistance to piperacillin, but full expression of cephalosporinase activity. Cefotaxime was a more discriminating drug than ceftazidime for detecting the presence of cephalosporinase, with a lower rate of hydrolysis of the latter in the presence of cephalosporinase. As shown in Table 7.13, phenotypes III and IV included strains that were cefotaxime-resistant, but only variably resistant to ceftazidime. The most threatening phenotype is phenotype V. Fortunately, strains exhibiting this profile are still isolated only rarely from nosocomial infections.

7.5.3 ß-Lactamase Inhibitors and *Acinetobacter* spp.

Of the available ß-lactamase inhibitors, clavulanic acid and sulbactam have been introduced and widely used in medical practice in combination with aminopenicillins or cephalosporins, and a newly developed compound, tazobactam, has also been investigated. ß-Lactamase inhibitors have poor intrinsic antibacterial activities, although sulbactam is intrinsically active against *Acinetobacter* spp. (Kitzis et al., 1983). Vallée et al. (1989) showed that sulbactam was less active (Table 7.14) than tazobactam against ticarcillin-resistant strains (MIC range of 1–32 mg/L), but both were active by themselves against ticarcillin-susceptible strains, with geometric mean MICs of sulbactam and tazobactam of 2.1 and 6.5 mg/L, respectively. It was of interest to examine the inhibitory activities of these ß-lactamase inhibitors against the ß-lactamases produced by *Acinetobacter* strains. MICs of ticarcillin, cefotaxime and ceftazidime were determined for these drugs alone and in combination with ß-lactamase inhibitors. Synergy was defined as a significant decrease in the geometric mean MICs of the antibiotic plus inhibitor compared with those for the antibiotic alone. Clavulanic acid caused a significant decrease in the MICs for the ticarcillin-resistant strains (MIC_{50} of 112 mg/L), but not a true reversion to susceptibility (Table 7.14). The MIC for ticarcillin-resistant strains (breakpoint of >64 mg/L) was diminished in combination with tazobactam to susceptible levels (geometric mean MIC of 33.02 mg/L). Similarly, sulbactam in

TABLE 7.14

Susceptibility of 46 Clinical Strains[a] and Two Susceptible Reference (ATCC) Strains of *A. calcoaceticus* to Ticarcillin in the Absence and Presence of β-Lactamase Inhibitors

Drugs		MICs (mg/L) Range	MIC_{50}	MIC_{90}	Geometric mean MIC
Ticarcillin	Tic^S	2 – 32	10	25.2	11.5
	Tic^R	2048 – >2048	>2048	>2048	>2048
Sulbactam	Tic^S	1 – 8	1.5	3.5	2.1
	Tic^R	16 – 32	16	22.7	18
Tazobactam	Tic^S	1 – 16	6	13.3	6.5
	Tic^R	1 – 32	5.6	14.8	7.1
Ticarcillin + clavulanic	Tic^S	0.5 – 32	9.7	22.9	10.6
acid (2mg/L)	Tic^R	32 – >2048	112	>2048	—
Ticarcillin + tazobactam	Tic^S	1 – 16	3.75	11.6	5.48
(ratio 1:1)	Tic^R	1 – 32	23.7	99.5	33.02
Ticarcillin + sulbactam	Tic^S	0.5 – 4	1.12	2.9	1.55
(ratio 1:1)	Tic^R	8 – 64	16	30.8	15.50
Ceftazidime		2 – 64	3.8	13.6	6.7
Cefotaxime		2 – 128	13.1	74.7	20.6
Ceftazidime	I[b]	—	—	—	1.0
+ Sulbactam	II	—	—	—	1.0
(ratio 1:1)	III	—	—	—	1.8 (5.2)
(+ Tazobactam)	IV	—	—	—	5.7 (3.6)

[a] 17 strains were ticarcillin-susceptible or moderately resistant; 29 strains were ticarcillin-resistant (MIC of >64 mg/L).

[b] Resistance phenotypes as listed in Table 7.13.

combination with ticarcillin reduced the MICs of ticarcillin to a geometric mean MIC of 15.50 mg/L. The geometric mean MICs of ceftazidime varied in relation to the enzymic profiles of the strains tested. Combinations of ceftazidime and inhibitors produced a significant reduction in the geometric mean MICs for strains that produced only a cephalosporinase, with the geometric mean MIC of ceftazidime reduced from 14.8 mg/L for phenotype III strains (cephalosporinase producers) to 1.8 and 5.2 mg/L when combined with sulbactam or tazobactam, respectively. A recent *in vitro* study (M.L. Joly-Guillou, unpublished results) has shown only limited bactericidal activity of ticarcillin, piperacillin and ceftazidime when used alone, but significant synergy with a combination of ticarcillin, clavulanic acid and sulbactam (256 + 2 + 8 mg/L). A decrease of 4–5 log cfu/ml within 24 h was observed with this non-conventional combination, which showed the best bactericidal effect against a cephalosporinase-producing strain. This results both from the intrinsic activity of sulbactam against

Acinetobacter, and the specific inhibitory power of ß-lactamase inhibitors against penicillinases and cephalosporinases. Tazobactam provided comparable activity, but was more limited than sulbactam; a combination of piperacillin plus tazobactam (128 + 4 mg/L) did not show any significant log cfu decrease, while the addition of sulbactam (8 mg/L) showed an initial moderate bactericidal effect (2 log cfu) with regrowth after incubation for 7–24 h.

7.5.4 Newer ß-Lactams

Recent *in vitro* studies have been carried out with newer cephalosporins, such as cefepime and cefpirome, and a new carbapenem, meropenem, in comparison with previously mentioned molecules (Thornsberry et al., 1993; Douboyas et al., 1994; Kuah et al., 1994; M.L. Joly-Guillou, unpublished results). The latter study tested 60 *A. baumannii* isolates against ceftazidime, cefpirome and imipenem; the MICs of cefpirome were slightly lower (geometric mean MIC of 3.6 mg/L) than those of ceftazidime (4.0 mg/L) for fully susceptible acinetobacters, while the MIC_{90}s for 25 strains of phenotype IV (penicillinase + cephalosporinase) were 128 and 64 mg/L with ceftazidime and cefpirome, respectively. Strains that produced only a cephalosporinase behaved variably in the presence of these drugs, but were predominantly resistant, with MICs ranging from 16–128 mg/L. Imipenem and meropenem inhibited most *Acinetobacter* strains, with MIC values of ≤0.5 mg/L in this study and others (Thornsberry et al., 1993; Kuah et al., 1994). Susceptible strains were inhibited with an MIC_{90} and MBC_{90} of ≤0.25 mg/L, confirming the potent bactericidal activity of imipenem. For cephalosporinase producers, the MIC_{90} of imipenem was slightly higher (1 mg/L), but for 10 strains of diminished susceptibility to imipenem the MIC_{90} was 4 mg/L, with a geometric mean MIC of 1.8 mg/L. Twenty-five non-*A. baumannii* strains were variably susceptible, with 60% of *A. haemolyticus*, *A. junii* and *A. lwoffii* isolates inhibited by ≤4 mg/L and ≤1 mg/L of cefpirome and imipenem, respectively.

7.5.5 Aminoglycosides

Mechanisms of resistance to aminoglycosides are described in detail in Chapter 8. All three types of aminoglycoside-modifying enzymes have been found in *Acinetobacter* spp. (Bergogne-Bérézin et al., 1980; Gomez-Lus et al., 1980; Shimizu et al., 1981; Devaud et al., 1982; Lambert et al., 1990; Bergogne-Bérézin et al., 1991). Recent surveys (Table 7.15) at Bichat Hospital in Paris, as well as a national survey in France (Joly-Guillou et al., 1991), have shown that 25–50% of strains were

TABLE 7.15

Resistance of *Acinetobacter* spp. to Aminoglycosides

Phenotypes	1984 – 1986[a]	1988 – 1990[a]	1991[b]
Susceptible (all aminoglycosides)	15[c]	9.5[c]	29[c]
Ka^R Ge^R	28	4	3.7
Ka^R Ge^R To^R Net^R	17.5	12	8.6
Ka^R Ge^R Am^R	15	12	<1
Ka^R Ge^R To^R Am^{R*}	—	4	18.5
Ka^R Net^R To^R Am^{R**}	—	8.5	<1
Resistant (all aminoglycosides)	24.5	50	27.2

Note: Ka, kanamycin; Ge, gentamicin; To, tobramycin; Net, netilmicin; Am, amikacin.

[a] Bichat-Claude Bernard Hospital.

[b] National survey (Joly-Guillou et al., 1992b); New phenotypes: * 1988; ** 1989.

[c] Expressed as a percentage.

resistant to all aminoglycosides, with variable rates of association between gentamicin, tobramycin and amikacin resistance. Two new phenotypes were identified in 1989 that combined resistance to tobramycin and amikacin. Most studies report 50–80% of isolates as being resistant to gentamicin and tobramycin, but 98% of isolates were reported to be resistant to aminoglycosides in a recent study from Germany (Seifert et al., 1993). Resistance rates were generally lower with netilmicin and amikacin, e.g., 72 and 66% of strains susceptible to amikacin and netilmicin, respectively (Vila et al., 1993).

It seems likely that aminoglycoside resistance in *Acinetobacter* may be more complex than in some other species; indeed, up to 74.5% of isolates in some series had multiple mechanisms of resistance. In France, the triple inactivating enzyme combination of AAD(2")-I + APH(3')-VI + AAC(3) is particularly common, while other combinations have included AAC(6')-I, AAC(3)-I, AAD(2")-I and APH(3')-VI. The latter enzyme, found at a high incidence, causes amikacin resistance (Joly-Guillou et al., 1987; Lambert et al., 1990). An international survey of 1189 aminoglycoside-resistant strains found that 34.8% of strains produced at least three enzymes, and various different combinations of enzymes have been observed (G. Miller, personal communication). Section 7.6 examines the use of aminoglycosides in combination with other antibiotics.

7.5.6 Fluoroquinolones

Since 1985, the new fluoroquinolones have been used increasingly for the treatment of severe nosocomial infections caused by multiply resistant bacteria, especially those compounds which can be administered parenterally. Table 7.16 compares the *in vitro* susceptibilities of *Acinetobacter* strains collected before and after the introduction of new quinolones into the antibacterial armamentarium. MIC values for recent strains were increased five- to tenfold for pefloxacin, ciprofloxacin and ofloxacin; furthermore, the MIC ranges were extended, e.g., from 0.016–32 mg/L to 0.03–128 mg/L for ciprofloxacin. The newer compounds (temafloxacin and sparfloxacin) did not show any significant improvement in activity compared to previous compounds, and the most active fluorinated quinolone against clinical isolates of *Acinetobacter* was ofloxacin. The proportions of fully susceptible strains were 27, 31 and 35% for pefloxacin, ciprofloxacin and ofloxacin, respectively. This threatening situation probably resulted from the early and large usage of pefloxacin in France. Another study (Chow et al., 1988), showed a more limited MIC range for ciprofloxacin of 0.5–4 mg/L, and MIC_{50} and MIC_{90} values of 2 and 4 mg/L, respectively, with a low inoculum of 10^4 cfu. The values were increased considerably to 8 and 16 mg/L, respectively, with a larger inoculum of 10^6 cfu, which suggests, as in the previous study, that there is a large proportion of resistant strains with MICs of >4 mg/L.

Figure 7.7 shows killing curves of *A. baumannii* obtained in the presence of four fluoroquinolones (Joly-Guillou and Bergogne-Bérézin, 1992). Sparfloxacin showed significantly better bactericidal activity than temafloxacin, while ciprofloxacin and pefloxacin were bacteriostatic only. The strains used in this study were more resistant than those found in some other studies (Table 7.16), but a recent study in Spain of 32 isolates (Garcia-Rodriguez et al., 1994) found similar MIC_{90} values of >16 mg/L for all fluoroquinolones tested. Possible mechanisms of resistance to fluoroquinolones are discussed in Chapter 8.

7.6 THERAPEUTIC APPLICATIONS OF ANTIBIOTIC COMBINATIONS AGAINST *Acinetobacter* spp.

7.6.1 Therapeutic Approaches

This chapter, as well as Chapters 5 and 6, has demonstrated the difficulties and failures in treating *Acinetobacter* infections over a period of more than 30 years as new antibiotics were introduced in turn for therapeutic use, either as single agents or more often in combinations. The recommended drugs in most recent studies have been either extended-spectrum penicillins, third-generation cephalosporins or imipenem, combined with an aminoglycoside.

TABLE 7.16

MICs of Fluoroquinolones for *A. baumannii*

Compound	Year	Range of MICs (mg/L)	MIC_{50} (mg/L)	MIC_{90} (mg/L)	Geometric mean MIC (mg/L)	Reference
Pefloxacin	1985	0.25 – 16	1.05	6.7	1.6	Joly-Guillou et al., 1992b
	1988 – 89	0.125 – 128	14.1	74.6	10.2	
	1990	0.125 – >128	46.3	196	24.3	
Ciprofloxacin	1985	0.016 – 32	0.7	4.6	0.98	
	1988 – 89	0.016 – 128	7.3	60.4	5.7	
	1990	0.03 –128	33.7	88.0	12.5	
	1991	0.03 – 1.0	0.06	0.5	—	Rolston et al., 1991
Ofloxacin	1985	0.06 – 16	0.69	3.7	1.03	Joly-Guillou et al., 1992b
	1988 – 89	0.016 – 32	3.37	14.6	2.03	
Temafloxacin	1988 – 89	0.016 – 64	2.8	41.0	2.8	
	1990	0.03 – 128	14.3	49.0	6.3	
	1991	0.03 – 1.0	0.06	0.12	—	Rolston et al., 1991
Sparfloxacin	1990	0.03 – 32	3.0	7.6	1.6	Joly-Guillou et al., 1992b
	1991	≤0.03 – 0.12	≤0.03	0.06	—	Rolston et al., 1991

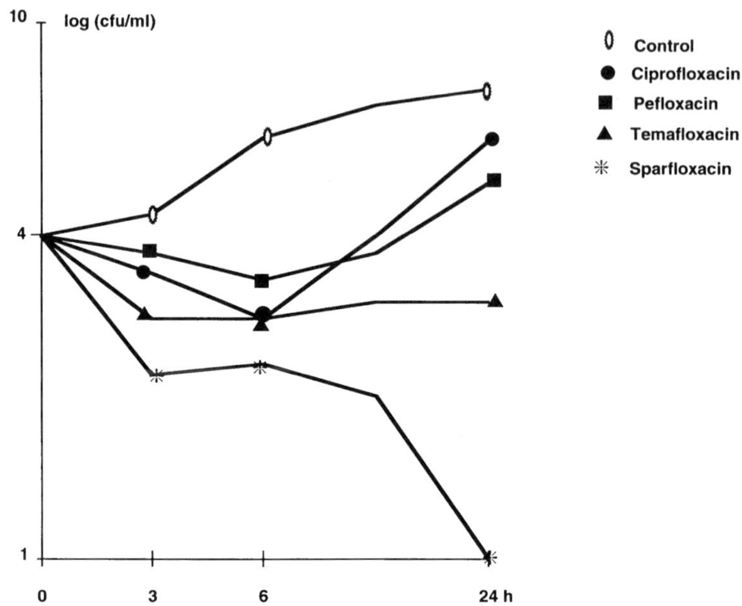

FIGURE 7.7
Killing curves of *A. baumannii* obtained in the presence of 4 × MIC of fluoroquinolones. (Data taken from Joly-Guillou et al., 1992b.)

In a retrospective survey in France of prescribing habits in 89 ICUs (Joly-Guillou et al., 1992b), the first-line therapy for *Acinetobacter* infections included amikacin, imipenem, ceftazidime and/or a quinolone (pefloxacin or ciprofloxacin); in 56% of cases, imipenem was prescribed either as a single agent or in combination with amikacin (18%), while ceftazidime plus amikacin was prescribed in 17% of cases, and amikacin was used as a single agent in 26% of cases. For second-line therapy after *in vitro* susceptibility testing, a combination of imipenem plus an aminoglycoside was used in 59% of cases, ceftazidime plus an aminoglycoside in 30% of cases, and ceftazidime combined with a quinolone in 11% of cases. A previous study in our hospital (Muller-Serieys et al., 1989) had shown that imipenem was used as monotherapy in 20% of 33 nosocomial infections, imipenem in combination with amikacin in 40%, and pefloxacin plus amikacin or tobramycin (depending on the antibiogram) in 20% of cases; treatment failure and death (caused by *Acinetobacter* infection and/or underlying disease) occurred in 17% of patients who received antibiotics. Many similar studies can be found in the medical literature (e.g., Vandenbrouck-Grauls et al., 1988; Buisson et al., 1990; Leonov et al., 1990; Amor et al., 1993).

7.6.2 *In Vitro* Activities of Antibiotic Combinations

Several recent studies of synergy and bactericidal activities of combinations have been carried out in order to provide an experimental basis for the choice of antibiotic combinations. A study of 20 *A. baumannii* strains (Xirouchaki and Giamarellou, 1992) compared combinations of aminoglycosides with imipenem or ciprofloxacin, with the bactericidal activities of the combinations evaluated by killing curves with concentrations of antibiotics similar to the serum levels achievable during treatment: i.e., amikacin 16 mg/L; other aminoglycosides 8 mg/L; imipenem 8 mg/L; and ciprofloxacin 1 mg/L. No significant difference between the tested combinations was found; when either imipenem or ciprofloxacin was combined with an aminoglycoside, enhanced killing was observed for 45–75% and 15–35%, respectively, of tested strains. A previous study (Chow et al., 1988) had shown that synergic interactions between ß-lactams or aminoglycosides plus ciprofloxacin were achieved less frequently than additive effects at clinically achievable serum levels. In another study, with highly resistant *Acinetobacter* isolates (Hercouet et al., 1989), combinations of imipenem with ticarcillin, or ticarcillin with netilmicin, were more rapidly bactericidal than imipenem or ticarcillin alone. Vila et al. (1993) described *in vitro* synergy of carbenicillin (similar to ticarcillin) with an aminoglycoside, even for strains only moderately susceptible to aminoglycosides. More recently, *in vitro* studies of bactericidal activities of ß-lactams combined with aminoglycosides have demonstrated enhanced killing of resistant *A. baumannii* isolates by combinations of ticarcillin + sulbactam + netilmicin (Joly-Guillou et al., 1995; Figures 7.8 and 7.9). Unusual combinations of piperacillin + tazobactam + sulbactam, or ticarcillin + clavulanic acid + sulbactam, although rather unconventional, may potentially have clinical relevance.

7.6.3 Newer Therapeutic Concepts for *Acinetobacter* Infections

Acinetobacter nosocomial infections are some of the most difficult-to-treat infections encountered by the clinician and may require combination therapy to achieve *in vivo* synergy. In severe infections, the size of the bacterial inoculum is important (Fantin and Carbon, 1992), since high bacterial counts offer more opportunities for the emergence of resistance. In experimental *Acinetobacter* pneumonia infections, the inoculum can be $\geq 10^8$ cfu/g of lung parenchyma (Joly-Guillou et al., 1994b), and is also high in clinical infections (Chastre et al., 1988). The use of antibiotic combinations may therefore help to prevent the emergence of resistance.

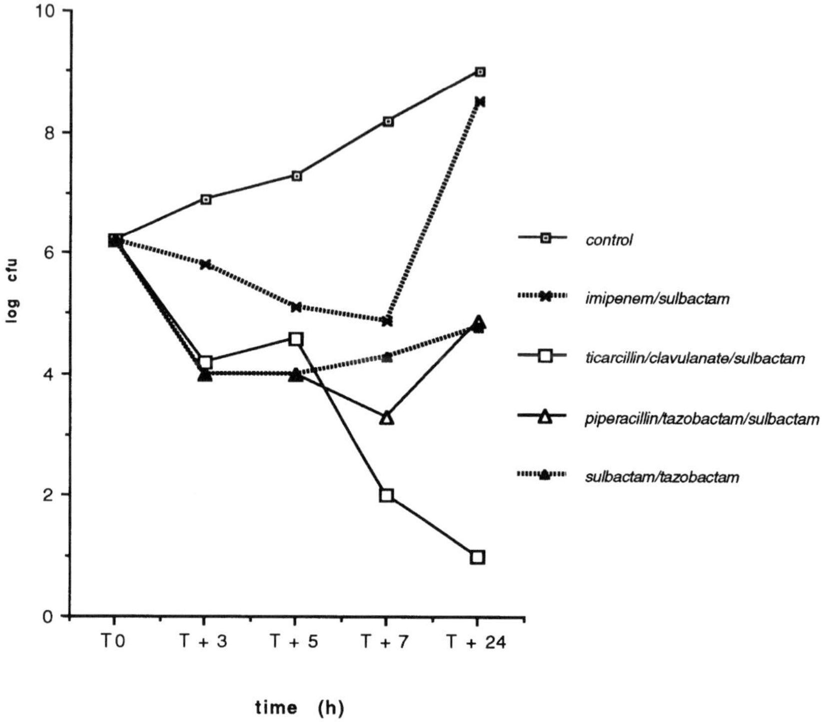

FIGURE 7.8
Activity of ß-lactams, with or without ß-lactamase inhibitors, against a cephalosporinase-producing strain of *A. baumannii*. (Data taken from Joly-Guillou et al., 1995.)

Another worrying parameter is the unpredictable correlation between *in vitro* synergy and the clinical outcome, e.g., with a combination including ciprofloxacin plus a ß-lactam (ceftazidime, ticarcillin, or imipenem). Combinations of aminoglycosides and fluoroquinolones have often shown indifferent or additive effects, both *in vitro* and in clinical trials (Moody et al., 1987; Neu, 1991). Ciprofloxacin plus imipenem is the combination used most frequently in our own hospital. *In vitro* killing curves of this combination demonstrate synergy and bactericidal activity against *Acinetobacter* isolates, and this combination has also resulted in favourable patient outcomes when treating *Acinetobacter* pneumonia.

The role of sulbactam, acting as a ß-lactamase inhibitor in association with its intrinsic antibacterial activity, in combinations with ß-lactams and/or an aminoglycoside should be defined and may offer a new therapeutic approach to problem *Acinetobacter* infections. The place

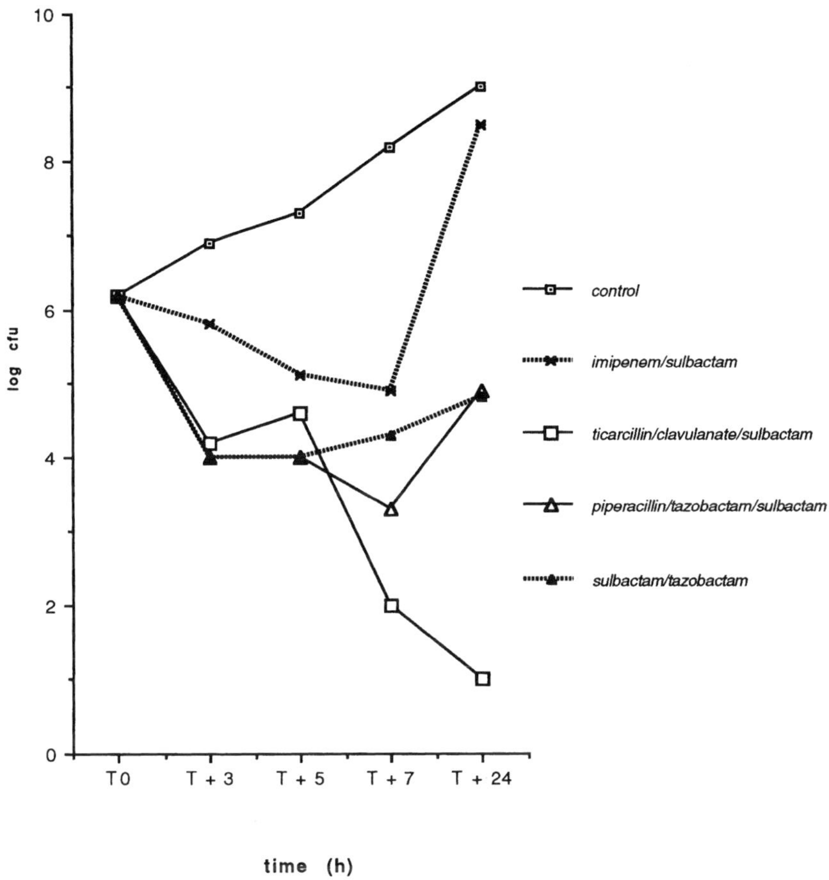

FIGURE 7.9
Activity of ß-lactams, with or without ß-lactamase inhibitors, against an imipenem-resistant (low-level) strain of *A. baumannii* producing both a cephalosporinase and a penicillinase. (Data taken from Joly-Guillou et al., 1995.)

of fluoroquinolones is more questionable because of their non-synergic activity in combinations, but newer molecules such as sparfloxacin seem to be more active and may provide improved potency when combined, after careful *in vitro* susceptibility testing, with ticarcillin, ceftazidime, or imipenem. The surprising activity of rifampicin against *Acinetobacter* (MIC range of 2–4 mg/L), as described by Bergogne-Bérézin et al. (1985), may also have a role, provided that rifampicin is included in a synergic combination in order to avoid or reduce the risk of spontaneous mutation to rifampicin resistance. Finally, prevention measures and antibiotic policies to control antibiotic use are also important measures for controlling *Acinetobacter* spread and contamination. In this respect, the possible role of selective digestive tract decontamination should also be defined (Joly-Guillou et al., 1994a; see also Chapter 4).

ACKNOWLEDGMENTS

I am extremely grateful to my colleagues (M.-L. Joly-Guillou, D. Decré, C. Muller-Serieys, E. Labourier) who have contributed for many years to studies with *Acinetobacter* spp. and also to Professor Vieu (Institut Pasteur) who helped our early attempts to understand the epidemiology of *Acinetobacter* and contributed to our typing studies for many years. Most of our work has been supported by grants from the Ministry of Health (Professor Jacques Roux), Roussel, Eli-Lilly, Glaxo and Merck-Sharp and Dohme Laboratories. Many thanks to Marie-Jeanne Julliard for her patient, efficient and excellent secretarial assistance.

REFERENCES

Acar, J., Bergogne-Bérézin, E., Brogard, J.M., Chabbert, Y., Cluzel, R., Courvalin, P., Dabernat, H., Drugeon, H., Dubreuil, L., Duval, J., Flandrois, J.P., Fleurette, J., Goldstein, F., Meyran, M., Morel, C., Philippon, A., Sirot, J., Soussy, C.J., Thabaut, A. and Veron, M. (1993). Communiqué 1993 du Comité de l'Antibiogramme de la Société Française de Microbiologie. *Pathol. Biol.* 41, 741–748.

Amor, A., Barguellil, F., Othmani, S. and Bahri, M. (1993). Infections à *Acinetobacter baumannii* et apport bactériologique. *Sem. Hôpitaux Paris* 69, 732–735.

Anstey, M.N., Currie, B.J. and Withrall K.M. (1992). Community-acquired *Acinetobacter* pneumonia in the Northern territory of Australia. *Clin. Infect. Dis.* 14, 83–91.

Bergogne-Bérézin, E. and Joly-Guillou, M.-L. (1985). An underestimated nosocomial pathogen, *Acinetobacter calcoaceticus*. *J. Antimicrob. Chemother.* 16, 535–538.

Bergogne-Bérézin, E. and Joly-Guillou, M.-L. (1986). Comparative activity of imipenem, ceftazidime and cefotaxime against *Acinetobacter calcoaceticus*. *J. Antimicrob. Chemother.* 18 (Suppl. E.), 35–39.

Bergogne-Bérézin, E. and Joly-Guillou, M.-L. (1991). Antibiotic resistance mechanisms in *Acinetobacter*. In *The Biology of* Acinetobacter, pp. 83–115. K.J. Towner, E. Bergogne-Bérézin and C.A. Fewson (Eds). Plenum Press, New York.

Bergogne-Bérézin, E., Zechovsky, N., Piéchaud, M., Vieu, J.F. and Bordini, A. (1971). Sensibilité aux antibiotiques de 240 souches de "*Moraxella*" oxydase-négative (*Acinetobacter*) isolées d'infections hospitalières. Evolution sur 3 ans. *Pathol. Biol.* 19, 981–990.

Bergogne-Bérézin, E., Joly, M.-L., Moreau, N. and Le Goffic, F. (1980). Aminoglycoside-modifying enzymes in clinical isolates of *Acinetobacter calcoaceticus*. *Curr. Microbiol.* 4, 361–364.

Bergogne-Bérézin, E., Joly-Guillou, M.-L. and Vieu, J.F. (1987). Epidemiology of nosocomial infections due to *Acinetobacter calcoaceticus*. *J. Hosp. Infect.* 10, 105–113.

Bousser, J., Hercouet, H., Thoreux, P.H., Camus, C., Michelet, C., Bouget, J., Thomas, R. and Cartier, F. (1988). Epidémie de pneumopathies nosocomiales à *Acinetobacter baumannii* dans un service de réanimation médicale. *Abstr. Réunion Interdisciplinaire de Chimiothérapie Antiinfectieuse (Paris).*

Bouvet, P.J.M. and Grimont, P.A.D. (1986). Taxonomy of the genus *Acinetobacter* with the recognition of *A. baumannii* sp. nov., *A. haemolyticus* sp. nov., *A. johnsonii* sp. nov., and *A. junii* sp. nov. and emended descriptions of *A. calcoaceticus* and *A. lwoffii*. *Int. J. Syst. Bacteriol.* 36, 228–240.

Buisson, Y., Tran van Nhieu, G., Ginot, L., Bouvet, P., Shill, H., Driot, L. and Meyran, M. (1990). Nosocomial outbreaks due to amikacin-resistant tobramycin-sensitive *Acinetobacter* species: correlation with amikacin usage. *J. Hosp. Infect.* 15, 83–93.

Chabbert, Y. and Courtieu, A.L. (1961). La sensibilité aux antibiotiques de *Pseudomonas*, *Achromobacter* et "BSW-*Bacterium anitratum*". *Ann. Inst. Pasteur* 100 (Suppl. 6), 100.

Chastre, J., Fagon, J.Y., Soler, P., Bornet, M., Domart, Y., Trouillet, J.L., Gibert, C. and Hance, A.J. (1988). Diagnosis of nosocomial bacterial pneumonia in intubated patients undergoing ventilation: comparison of the usefulness of bronchoalveolar lavage and the protected specimen brush. *Am. J. Med.* 85, 499–506.

Chopade, B.A., Wise, P.J. and Towner, K.J. (1985). Plasmid transfer and behaviour in *Acinetobacter calcoaceticus* EBF65/65. *J. Gen. Microbiol.* 131, 2805–2811.

Chow, A.W., Wong, J. and Bartlett, K.H. (1988). Synergistic interactions of ciprofloxacin and extended spectrum beta-lactams or aminoglycosides against *Acinetobacter calcoaceticus* ss. *anitratus*. *Diagn. Microbiol. Infect. Dis.* 9, 213–217.

Cordes, L.G., Brink, E.W. and Checko P.J. (1981). A cluster of *Acinetobacter* pneumonia in foundry workers. *Ann. Intern. Med.* 95, 688–695.

Courcol, R.J., Pinkas, M. and Martin, G.R. (1989). A seven-year survey of antibiotic susceptibility and its relationship with usage. *J. Antimicrob. Chemother.* 23, 441–451.

Courtieu, A.L. and Chassignol, S. (1961). *Les Germes du Groupe* Moraxella (Acinetobacter), Bacterium anitratum. *Fréquence. Bactériologie.* Fayolle, Lyon.

Courvalin, P. (1994). Transfer of antibiotic resistance genes between Gram-positive and Gram-negative bacteria. *Antimicrob. Agents Chemother.* 38, 1447–1451.

Devaud, M., Kayser, F.H. and Bachi, B. (1982). Transposon-mediated multiple antibiotic resistance in *Acinetobacter* strains. *Antimicrob. Agents Chemother.* 22, 323–329.

Douboyas, J., Tzouvelekis, L.S. and Tsakris, A. (1994). *In vitro* activity of ampicillin/sulbactam against multiresistant *Acinetobacter calcaoceticus* var. *anitratus* clinical isolates. *J. Antimicrob. Chemother.* 34, 298–300.

Dowding, J.E. (1979). A novel aminoglycoside-modifying enzyme from a clinical isolate of *Acinetobacter*. *J. Gen. Microbiol.* 110, 239–241.

Fantin, B. and Carbon, C. (1992). *In vivo* antibiotic synergism: contribution of animal models. *Antimicrob. Agents Chemother.* 26, 1–4.

Finland, M. (1979). Emergence of antibiotic resistance in hospitals, 1935–1975. *Rev. Infect. Dis.* 1, 4–21.

French, G.L., Casewell, M.W., Roncoroni, A.J., Knight, S. and Phillips, I. (1980). A hospital outbreak of antibiotic-resistant *Acinetobacter anitratus*: epidemiology and control. *J. Hosp. Infect.* 1, 125–131.

Freney, J., Bouvet, P.J.M. and Tixier, C. (1989). Identification et détermination de la sensibilité aux antibiotiques de 31 souches cliniques d'Acinetobacter autres que *A. baumannii*. *Ann. Biol. Clin.* 47, 41–44.

Garcia, I., Fainstein, V., Leblanc, B. and Bodey, G.P. (1983). *In vitro* activities of new beta-lactam antibiotics against *Acinetobacter* spp. *Antimicrob. Agents Chemother.* 24, 297–299.

Garcia-Rodriguez, J.A., Garcia-Sanchez, J.E., Garcia-Garcia, M.I., Fresnadillo, M.J., Trujillano, I. and Garcia-Sanchez, E. (1994). *In vitro* activity of four new fluoroquinolones. *J. Antimicrob. Chemother.* 34, 53–64.

Gardner, P., Smith, D.H., Beer, H. and Moellering, R.C. (1969). Recovery of resistance (R) factors from a drug-free community. *Lancet* ii, 774–776.

Gehrlein, M., Leying, H., Cullmann, W., Wendt, S. and Opferkuch, W. (1991). Imipenem resistance in *Acinetobacter baumannii* is due to altered penicillin-binding proteins. *Chemotherapy* 37, 405–412.

Gerner-Smidt, P. (1987). The epidemiology of *Acinetobacter calcoaceticus*: biotype and resistance-pattern of 328 strains consecutively isolated from clinical specimens. *Acta Pathol. Microbiol. Immunol. Scand.* 95B, 5–11.

Go, S.E., Urban, C., Burns, J., Kreiswirth, B., Eisner, W., Mariano, N., Mosinka-Snipas, K. and Rahal, J.J. (1994). Clinical and molecular epidemiology of acinetobacter infections sensitive only to polymyxin B and sulbactam. *Lancet* 344, 1329–1332.

Godineau-Gauthey, N., Lesage, D., Tessier, F., Kollia, D. and Daguet, G.L. (1988). *Acinetobacter calcoaceticus* variété *anitratus* ou *Acinetobacter baumannii*. Etude de la sensibilité aux antibiotiques chez 65 souches hospitalières. *Méd. Mal. Infect.* ii, 124.

Goldstein, G.W., Labigne-Roussel, A., Gerbaud, G., Carlier, C., Collatz, E. and Courvalin, P. (1983). Transferable plasmid mediated antibiotic resistance in *Acinetobacter*. *Plasmid* 10, 138–147.

Gomez-Lus, R., Larrad, L., Rubio-Calvo, M.C., Navarro, M. and Asierra, M.P. (1980). AAC(3) and AAC(6') enzymes produced by R plasmids isolated in general hospital. In *Antibiotic Resistance*, pp. 295–303. S. Mitsuhashi, L. Rosival and V. Kremery (Eds). Springer-Verlag, Berlin.

Goodhart, G.L., Abrutyn, E., Watson, R., Root, R.K. and Egert, J. (1977). Community- acquired *Acinetobacter calcoaceticus* var. *anitratus* pneumonia. *J. Am. Med. Assoc.* 238, 1516–1518.

Gould, I.M. (1994). Risk factors for acquisition of multiply drug resistant Gram negative bacteria. *Eur. J. Clin. Microbiol. Infect. Dis.* 13 (Suppl. 1), 30–38.

Gradon, J.D., Chapnick, E.K. and Lutwick, L.I. (1992). Infective endocarditis of a native valve due to *Acinetobacter:* case report and review. *Clin. Infect. Dis.* 14, 1145–1148.

Heisig, P. and Wiedemann B. (1991). Use of broad host range *gyrA* plasmid for genetic characterization of fluoroquinolone resistant Gram-negative bacteria. *Antimicrob. Agents Chemother.* 35, 2031–2036.

Hercouet, H., Bousser, J., Donnio, P.Y. and Avril, J.L. (1989). Activité *in vitro* des antibiotiques sur les souches hospitalières de *Acinetobacter baumannii. Pathol. Biol.* 37, 612–616.

Hoffmann, S., Mabeck, C.E. and Vejlsgaard, R. (1982). Bacteriuria caused by *Acinetobacter calcoaceticus* biovars in a normal population and in general practice. *J. Clin. Microbiol.* 16, 443–451.

Joly-Guillou, M.-L. and Bergogne-Bérézin, E. (1985). Evolution d'*Acinetobacter calcoaceticus*, en milieu hospitalier, de 1971 à 1984. *Presse Méd.* 14, 2331–2335.

Joly-Guillou, M.-L. and Bergogne-Bérézin, E. (1992). *In vitro* activity of sparfloxacin, pefloxacin, ciprofloxacin and temafloxacin against clinical isolates of *Acinetobacter* spp. *J. Antimicrob. Chemother.* 29, 466–468.

Joly-Guillou, M.-L., Bergogne-Bérézin, E. and Moreau, N. (1987). Enzymatic resistance to beta-lactams and aminoglycosides in *Acinetobacter calcoaceticus. J. Antimicrob. Chemother.* 20, 773–776.

Joly-Guillou, M.-L., Vallée, E., Bergogne-Bérézin, E. and Philippon, A. (1988). Distribution of beta-lactamases and phenotype analysis in clinical strains of *Acinetobacter calcoaceticus. J. Antimicrob. Chemother.* 22, 597–604.

Joly-Guillou, M.-L., Bergogne-Bérézin, E. and Vieu, J.F. (1990). Epidémiologie et résistance aux antibiotiques des *Acinetobacter* en milieu hospitalier. Bilan de 5 années. *Presse Méd.* 19, 357–361.

Joly-Guillou, M.-L., Decré, D. and Bergogne-Bérézin, E. (1992a). Infections nosocomiales à *Acinetobacter*: Surveillance épidémiologique hospitalière. *Bull. Epidémiol. Hebdominaire.* vL, 211–212.

Joly-Guillou, M.-L., Decré, D., Wolff, M. and Bergogne-Bérézin, E. (1992b). *Acinetobacter* spp: clinical epidemiology in 89 intensive care units. A retrospective study in France during 1991. *Abstr. 2nd Int. Conf. Prevention of Infection (Nice).*

Joly-Guillou, M.-L., Decré, D., Desjardin, P., Bingen, E., Lambert-Zechovsky, N. and Bergogne-Bérézin, E. (1992c). Résistance à l'imipenem chez *Acinetobacter baumannii* producteur d'une nouvelle ß-lactamase: analyse moléculaire de deux souches isolées en 1986 et 1988 dans le CHU Bichat. *Abstr. Réunion Interdisciplinaire de Chimiothér. Anti-Infect. (Paris).*

Joly-Guillou, M.-L., Wolff, M., Decré, D., Calvat, S. and Bergogne-Bérézin, E. (1994a). Colonisation and infection with *Acinetobacter baumannii* in ICU patients receiving selective digestive decontamination. Results of a case-control and molecular epidemiologic investigations. *Abstr. 34th Intersci. Conf. Antimicrob. Agents and Chemother. (Orlando).*

Joly-Guillou, M.-L., Wolff, M., Walker F. and Vallée, E. (1994b). A mouse model of *Acinetobacter baumannii* pneumonia. *Abstr. 3rd Int. Symp. Biology of* Acinetobacter *(Edinburgh).*

Joly-Guillou, M.-L., Decré, D., Herrman, J.L., Bourderlier, E. and Bergogne-Bérézin, E. (1995). Bactericidal *in vitro* activity of ß-lactams and ß-lactamase inhibitors, alone or associated, against clinical strains of *Acinetobacter baumannii*: effect of combination with aminoglycosides. *J. Antimicrob. Chemother.*, 36, in press.

Kitzis, M.D., Goldstein, F.W., Labia, R. and Acar, J.F. (1983). Activité du sulbactam et de l'acide clavulanique, seuls et associés, sur *Acinetobacter calcoaceticus. Ann. Microbiol. Inst. Pasteur* 134A, 163–168.

Kuah, B.G., Kumarasinghe, G., Doran, H. and Chang, H.R. (1994). Antimicrobial susceptibilities of clinical isolates of *Acinetobacter baumannii* from Singapore. *Antimicrob. Agents Chemother.* 38, 2502–2503.

Lambert, T., Gerbaud, G., Bouvet, P., Vieu, J.F. and Courvalin, P. (1990). Dissemination of amikacin resistance gene *aphA6* in *Acinetobacter* spp. *Antimicrob. Agents Chemother.* 34, 1244–1248.

Larson, T. (1984). A decade of nosocomial *Acinetobacter. Am. J. Infect. Control* 12, 14–18.

Leonov, Y., Schlaeffer, F., Karpuch, J., Bourvin, A., Shemesh, Y. and Lewinson, G. (1990). Ciprofloxacin in the treatment of nosocomial multiply resistant *Acinetobacter calcoaceticus* bacteremia. *Infection* 18, 234–236.

Linton, A.H. (1986). Flow of resistance genes in the environment and from animals to man. *J. Antimicrob. Chemother.* 18 (Suppl. C), 189–197.

McGowan, J.E. (1983). Antimicrobial resistance in hospital organisms and its relation to antibiotic use. *Rev. Infect. Dis.* 5, 1033–1048.

Moody, J.A., Gerding, D.N. and Peterson, L.R. (1987). Evaluation of ciprofloxacin's synergism with other agents by multiple *in vitro* methods. *Am. J. Med.* 82, 44–54.

Morohoshi, T. and Saito, T. (1977). Beta-lactamase and beta-lactam antibiotic resistance in *Acinetobacter anitratus* (syn. *A. calcoaceticus*). *J. Antibiot.* 30, 969–973.

Muller-Serieys, C., Lesquoy, J.B., Perez, E., Fichelle, A., Boujeois, B., Joly-Guillou, M.-L. and Bergogne-Bérézin, E. (1989). Infections nosocomiales à *Acinetobacter*: épidémiologie et difficultés thérapeutiques. *Presse Méd.* 18, 107–110.

Murray, B.E. and Moellering, R.C. (1980). Evidence of plasmid mediated production of aminoglycoside-modifying enzyme not previously described in acinetobacter. *Antimicrob. Agents Chemother.* 17, 30–36.

Neu, H.C. (1991). Synergy and antagonism of combinations with quinolones. *Eur. J. Clin. Microbiol. Infect. Dis.* 10, 255–261.

Obana, Y., Nishino, T. and Tanino, T. (1985). *In vitro* and *in vivo* activities of antimicrobial agents against *Acinetobacter calcoaceticus*. *J. Antimicrob. Chemother.* 15, 441–448.

Obara, M. and Nakai, T. (1991). Mechanisms of resistance to ß-lactam antibiotics in *Acinetobacter calcoaceticus*. *J. Antimicrob. Chemother.* 28, 791–800.

Paton, R., Miles, R.S., Hood, J. and Amyes, S.G.B. (1993). ARI-1: ß-lactamase-mediated imipenem resistance in *Acinetobacter baumannii*. *Int. J. Antimicrob. Agents.* 2, 81–88.

Paul, G., Joly-Guillou, M.-L., Bergogne-Bérézin, E., Nevot, P. and Philippon A. (1989). Novel carbenicillin-hydrolyzing beta-lactamase (CARB-5) from *Acinetobacter calcoaceticus* var. *anitratus*. *FEMS Microbiol. Lett.* 59, 45–50.

Phillips, I., King, A., Shannon, K. and Warren, C. (1981). SQ 26776: *in vitro* antibacterial activity and susceptibility to ß-lactamases. *J. Antimicrob. Chemother.* 8 (Suppl. E), 103–107.

Piddock, L.J.V. and Wise, R. (1985). Mechanisms of resistance to quinolones and clinical perspectives. *J. Antimicrob. Chemother.* 23, 475–479.

Pinter, M. and Kantor, M. (1971). Quantitative antibiotic sensitivity pattern of *Acinetobacter* strains. *Antonie van Leeuwenhoek J. Microbiol.* 37, 197.

Ramphal, R. and Kluge, R.M. (1979). *Acinetobacter calcoaceticus* variety *anitratus*: an increasing nosocomial problem. *Am. J. Med. Sci.* 277, 57–66.

Retailliau, H.F., Hightower, A.W., Dixon, R.E. and Allen, J.R. (1979). *Acinetobacter calcoaceticus*: a nosocomial pathogen with an unusual seasonal pattern. *J. Infect. Dis.* 139, 371–375.

Rolston, K., Messer, M., Nguyen, H., Ho, D.H., LeBlanc, B. and Bodey, G.P. (1991). Susceptibility of *Pseudomonas* spp., *Xanthomonas maltophilia*, *Acinetobacter* spp. and *Alcaligenes denitrificans* subsp. *xylosoxidans* to newer quinolones. *Eur. J. Clin. Microbiol. Infect. Dis. Special Issue (Proc. 3rd Int. Symp. New Quinolones, Vancouver)*, 679–683.

Rudin, M.L., Michael, J.R. and Huxley, E.J. (1979). Community-acquired *Acinetobacter* pneumonia. *Am. J. Med.* 67, 39–43.

Santos-Ferreira, M.O., Vieu, J.F. and Klein, B. (1984). Phage-types and susceptibility to 26 antibiotics of nosocomial strains of *Acinetobacter* isolated in Portugal. *J. Intern. Med. Res.* 12, 364–368.

Sato, K. and Nakae, T. (1991). Outer membrane permeability of *Acinetobacter calcoaceticus* and its implication in antibiotic resistance. *J. Antimicrob. Chemother.* 28, 35–45.

Seifert, H., Baginski, R., Schulze, A. and Pulverer, G. (1993). Antimicrobial susceptibility of *Acinetobacter* species. *Antimicrob. Agents Chemother.* 37, 750–753.

Shimizu, S., Inoue, M., and Mitsuhashi, S. (1981). Enzymatic adenylation of spectinomycin by *Acinetobacter calcoaceticus* subsp. *anitratus*. *J. Antibiot.* 34, 869–875.

Struelens, M.J., Carlier, E., Maes, N., Serruys, E., Quint, W.V.G., and van Belkum, A. (1993). Nosocomial colonization and infection with multiresistant *Acinetobacter baumannii*: outbreak delineation using DNA macrorestriction analysis and PCR-fingerprinting. *J. Hosp. Infect.* 25, 15–32.

Tankovic, K, Legrand, P., De Gatines, G., Chemineau, V., Brun-Buisson, C. and Duval, J. (1994). Characterization of a hospital outbreak of imipenem-resistant *Acinetobacter baumannii* by phenotypic and genotypic typing methods. *J. Clin. Microbiol.* 32, 2677–2681.

Thornsberry, C., Marler, J.K., Brown, S.D., Yee, C., Bouchillon, S.K. and Rich, T. (1993). Clinical acinetobacters: which species? what susceptibility? *Abstr. 33rd Intersci. Conf. Antimicrob. Agents and Chemother. (New Orleans).*

Towner, K.J. (1991). Plasmid and transposon behaviour in *Acinetobacter*. In *The Biology of* Acinetobacter, pp. 149–167. K.J. Towner, E. Bergogne-Bérézin and C.A. Fewson (Eds). Plenum Press, New York.

Traub, W.H. and Spohr, M. (1989). Antimicrobial drug susceptibility of clinical isolates of *Acinetobacter* species (*A. baumannii*, *A. haemolyticus*, genospecies 3, and genospecies 6). *Antimicrob. Agents Chemother.* 33, 1617–1619.

Urban, C., Go, E., Mariano, N., Berger, B.J., Avraham, I., Rubin, D. and Rahal, J.J. (1993). Effect of sulbactam on infections caused by imipenem-resistant *Acinetobacter calcoaceticus* biotype *anitratus*. *J. Infect. Dis.* 167, 448–451.

Vallée, E., Joly-Guillou, M.-L. and Bergogne-Bérézin, E. (1989). *Acinetobacter calcoaceticus*. Resistance aux beta-lactamines et beta-lactamases. *Med. Mal. Infect. Hors-série Mai*, 65–71.

Vandenbroucke-Grauls, C.M.J.E., Kerver, A.J.H., Rommes, J.H., Jansen, R., den Dekker, C. and Verhoef, J. (1988). Endemic *Acinetobacter anitratus* in a surgical intensive care unit: mechanical ventilators as reservoir. *Eur. J. Clin. Microbiol. Infect. Dis.* 7, 485–489.

Vila, J., Marcos, A., Marco, F., Abdalla, S., Bergara, Y., Reig, R., Gomez-Lus, R. and Jimenez de Anta, T. (1993). *In vitro* antimicrobial production of ß-lactamases, aminoglycoside-modifying enzymes, and chloramphenicol acetyltransferase by and susceptibility of clinical isolates of *Acinetobacter baumannii*. *Antimicrob. Agents Chemother.* 37, 138–141.

Xirouchaki, E. and Giamarellou, H. (1992). *In vitro* interactions with imipenem or ciprofloxacin against aminoglycoside resistant *Acinetobacter baumannii*. *J. Chemother.* 4, 263–267.

Chapter 8

Mechanisms of Antibiotic Resistance in *Acinetobacter* spp. — Genetics of Resistance

S.G.B. Amyes and H.-K. Young

CONTENTS

0-8493-9223-3/95/$0.00+$.50

8.1 INTRODUCTION

As described in the previous chapter, multiple antibiotic resistance has been a feature of *Acinetobacter* spp. ever since attention was first drawn to their potential role as significant human pathogens (Bergogne-Bérézin et al., 1971). Perhaps surprisingly, some multiresistant strains were isolated from surgical patients before any obvious antibiotic therapy had been administered, and it appears that *Acinetobacter* is a genus which has a propensity to develop antibiotic resistance extremely rapidly, perhaps as a consequence of long-term evolutionary exposure of the genus to antibiotic-producing organisms in a soil environment. This is in contrast to more 'traditional' clinical bacteria which seem to require more time to acquire highly effective resistance mechanisms in response to the introduction of modern radical therapeutic strategies.

If it is accepted that acinetobacters are relatively rare pathogens in an antibiotic-free environment, then it is probably their ability to respond rapidly to challenge with antibiotics, coupled with widespread use of antibiotics in the hospital environment, that is responsible for their success as nosocomial pathogens. This chapter reviews the current state of knowledge regarding antibiotic resistance mechanisms in *Acinetobacter* spp. and summarises the problems that may be posed in the future.

8.2 β-LACTAM RESISTANCE

8.2.1 Plasmid-Associated ß-Lactamases

The role of ß-lactamases in penicillin and cephalosporin resistance in *Acinetobacter* spp. is relatively straightforward. The high incidence of ß-lactam resistance found in the early strains and the ever-increasing proportion of resistant isolates suggested that plasmid-associated ß-lactamases should be prevalent. Reports of ß-lactamases in *Acinetobacter* spp. started with identification of the ubiquitous class A TEM-1 ß-lactamase in epidemic strains in the early 1980s (Philippon et al., 1980; Goldstein et al., 1983), followed by the closely related TEM-2 ß-lactamase (Devaud et al., 1982). The ampicillin resistance of the host strain was attributed to the presence of these enzymes. However, when 76 ticarcillin-resistant (MIC of >256 mg/L) *Acinetobacter* strains were analysed for their ß-lactamase content by isoelectric separation according to their pI by the method of Matthew et al. (1975), with visualisation of the proteins by the iodometric method with benzylpenicillin as an indicator, penicillinase activity was found in only 41% of the resistant strains (Joly-Guillou et al., 1988). The majority of these strains produced an enzyme with a pI of 5.4 (TEM-1-like), although a few had an enzyme with a pI of 6.3, which is characteristic of the ß-lactamase CARB-5. This serine active-site class A ß-lactamase hydrolyses both carbenicillin and ampicillin; however, it gives poor hydrolysis of cephalosporins and virtually none of methicillin. The enzyme is susceptible to inhibition by *p*-chloromercuribenzoate (*p*CMB), which demonstrates the presence of cysteine residues (Bauernfeind, 1986), and to the ß-lactamase inhibitors clavulanic acid and sulbactam, suggesting that it belongs to class A (Ambler, 1980). The substrate profile suggests that CARB-5 is a penicillinase, and its high activity against carbenicillin as a substrate places it as a carbenicillinase (Paul et al., 1989). However, the presence of these two enzymes could not explain all the ß-lactam resistance in the *Acinetobacter* strains studied. Some ß-lactamase activity was also identified with a pI above 8.0. These were undefined enzymes which were presumed to be chromosomally encoded cephalosporinases because of their high pI.

A separate study in France isolated a strain of *A. baumannii* from a urine sample in late 1989 that exhibited an unusual susceptibility profile with resistance to ticarcillin and piperacillin (MICs of >256 mg/L), cefoxitin (64 mg/L), cefotaxime (32 mg/L) and ceftazidime (16 mg/L) (Joly-Guillou and Bergogne-Bérézin, 1990). Analysis of crude ß-lactamase extracts by the gel iodometric method in the presence of substrate inhibitors showed that the ß-lactamase activity present was inhibited by clavulanic acid (which specifically inhibits penicillinase activity), but not by cloxacillin (which specifically inhibits cephalosporinase activity). Analytical isoelectric focusing (IEF) failed to detect either a TEM-type or a CARB-5 enzyme, or any significant cephalosporinase activity, but some enzymic activity was detected at a pI of 7.7. Although it was postulated that this particular *Acinetobacter* isolate may have acquired the ability to produce one of the 'broad-spectrum' ß-lactamases described by Sirot et al. (1988), some of which are derived from the SHV ß-lactamase family and have a pI of about 7.7, this seems unlikely as all the SHV-derived broad-spectrum ß-lactamases have significant hydrolytic activity against cefotaxime and ceftazidime (Du Bois et al., 1995), but this enzyme has none.

In a later study by Vila et al. (1993), antimicrobial susceptibility testing was performed on 54 epidemiologically unrelated clinical isolates of *A. baumannii*. Cephalosporinase activity was found in 98% of the strains. This was based on IEF bands with a pI above 8.0. An enzyme with a pI of 5.4, presumed to be the TEM-1 ß-lactamase, was found in 16% of the strains. However, only 2% of the strains were susceptible to amoxycillin at the standard breakpoint (National Committee for Clinical Laboratory Standards, 1990), with an MIC_{50} of 64 mg/L and MIC_{90} of >256 mg/L. Even when amoxycillin was combined with clavulanic acid, only 16% were considered sensitive at the standard breakpoint (National Committee for Clinical Laboratory Standards, 1990), with the MIC_{50} dropping only to 16 mg/L and the MIC_{90} to 32 mg/L. These results are not commensurate with the presence of TEM-1 ß-lactamase, and no ß-lactamase activity was found at pI 6.3 indicative of CARB-5. However, 11% of the strains had a ß-lactamase with a pI between 7.0 and 7.5. The properties of this enzyme were not determined, nor was it identified; however, even its presence could not explain the extent of ß-lactam resistance. These results suggest that cephalosporinases are the predominant ß-lactamases.

8.2.2 Chromosomal ß-Lactamases

By comparison with other Gram-negative genera, it might be speculated that the main harbingers of ß-lactam resistance in *Acinetobacter* spp. are ß-lactamases that were probably introduced on bacterial plasmids. However, there are a number of Gram-negative bacteria (e.g., the *Klebsiella* genus) that are able to either induce or produce chromosomal cephalosporinases constitutively. This capability has ensured that these species are considered to be ampicillin-resistant, even though a conventional sensitivity test may not indicate this. Unlike *Escherichia coli, Klebsiella* spp. are common environmental bacteria as well as clinical pathogens and, as such, appear to have developed a ready capability to fend off attacks by environmentally produced ß-lactams. *Acinetobacter* spp. are also environmental bacteria turned clinical pathogens, and thus might be expected to have the same capability.

The presence of chromosomal ß-lactamases in the genus *Acinetobacter* was first suggested by Matthew and Harris (1976), who demonstrated ß-lactamases with pI values around 8.6, and this was reiterated by

Sykes and Matthew (1976). At this time there was considerable confusion about the classification of the genus, and these authors studied only one strain of *Acinetobacter*, termed *Acinetobacter inotti*. Subsequent analysis of this strain by Hood and Amyes (1991) suggests that it was actually *A. lwoffii*. This appears to be the first true chromosomal *Acinetobacter* ß-lactamase reported. In the following decade, ß-lactamases with pIs ranging from 7.5 to >10 were reported (Medeiros et al., 1985; Joly-Guillou et al., 1987; Hikida et al., 1989). Medeiros et al. (1985) noted that some of these enzymes focused poorly with conventional polyacrylamide isoelectric focusing and, in a study of *Acinetobacter* strains, determined the pIs of three enzymes as 8.8, 9.0 and 9.7, respectively, while four enzymes had a pI of exactly 10.0, and three had a pI of >10. Most significantly, four enzymes failed to focus as discrete bands, but formed a blur. Examination of the ß-lactamases from 30 strains of *Acinetobacter* (Joly-Guillou et al., 1987) revealed a cephalosporinase with a pI >8.0. This suggests that the enzymes failed to focus as discrete bands on polyacrylamide IEF gels. Whether Hikida et al. (1989) had difficulty in focusing their *Acinetobacter* ß-lactamases is not recorded, but they reported that the purified chromosomal ß-lactamase of their strain ML4961 had a pI of 9.9. However, a broad range (pH 3.5–10) ampholine was used in their IEF gel, and it is therefore likely that their enzyme had migrated to the cathode.

The difficulty in distinguishing these ß-lactamases by IEF has left some uncertainty about the number of ß-lactamases present in a particular strain. Thus, if the properties of ß-lactamases have been studied from strains producing more than one enzyme and these have not been separated, then some difficulty could be experienced in interpretation of the results. Morohoshi and Saito (1977) examined the hydrolysis of various ß-lactams by the ß-lactamases of *Acinetobacter* spp., and concluded that the main enzyme activity was a cephalosporinase, with an M_r of 30,000. The predominance of cephalosporinase activity was supported by Hikida et al. (1989) who examined the susceptibility of an *Acinetobacter* ß-lactamase to inhibitors. However, their enzyme had an M_r of 38,000. On the basis of the data of Morohoshi and Saito (1977), the *Acinetobacter* ß-lactamase was classified in Group 1 (Bush, 1989); i.e., a cephalosporinase that was not inhibited by clavulanic acid (CEP-N). This classification was supported by the subsequent substrate and inhibition studies of Hikida et al. (1989).

Eight ß-lactamases were partially purified from aztreonam-resistant strains collected in the Royal Infirmary Edinburgh (Table 8.1) by Hood and Amyes (1989). When these ß-lactamases were applied to a conventional polyacrylamide IEF system, seven did not focus. The ß-lactamase from one strain, previously identified as *A. lwoffii*, but now reclassified as *A. junii*, focused at a pI of 9.1. These enzymes were also

TABLE 8.1

Source of the ACE Chromosomal ß-Lactamases

Strain No.	Species	Specimen source	AZT MIC (mg/L)	ACE enzyme
6B230	*A. junii*	Blood	32	1
H17	*A. baumannii*	Urine	64	1
H26	*A. baumannii*	Sinus	32	1
H68	*A. baumannii*	Urine	64	1
H141	*A. baumannii*	Wound	64	1
H162	*A. baumannii*	Wound	64	2
H126	*A. junii*	Wound	>256	3
5B86	*A. lwoffii*	Blood	16	4

Note: AZT = aztreonam.

difficult to separate on another polyacrylamide system, SDS-free PAGE. After separation and visualisation with nitrocephin, all the enzyme activity had migrated to the anode, suggesting that the enzymes were migrating on charge and not, as they should in conventional PAGE, by molecular mass. This implied that the conventional IEF system employing polyacrylamide was failing because the enzymes were not entering the gel matrix. In other words they lacked sufficient solubility. The enzymes could be rendered soluble if SDS was present, but this ran the risk of destroying enzyme activity and making detection almost impossible.

A new approach was attempted to try and solubilise the proteins without a severe effect on activity. This required a less devastating denaturation of the protein than that provided by SDS. Initially, 4 *M* urea and Triton® X-100 2.5% were incorporated into the polyacrylamide gel matrix. However, only when the enzyme samples were solubilised with urea, Triton® X-100, ß-mercaptoethanol and ampholines was focusing possible and visible with nitrocephin staining. Although separation was possible, some molecular sieving was occurring, causing distortion of the bands. This resulted from the relatively small pore size of the polyacrylamide matrix. Vecoli et al. (1983) used agarose IEF for the investigation of ß-lactamases, and Saravis and Zamcheck (1979) described thin layer IEF gels containing agarose which resolved proteins of M_r >2 × 10^6. However, the incorporation of urea into agarose gels lowers their mechanical strength and risks carbamylation of the proteins being studied (Olsson and Laas, 1981). These authors incorporated sorbitol and ß-mercaptoethanol into their gels to overcome this problem. Therefore, gels containing sorbitol and ß-mercaptoethanol, agarose 1%, Triton® X-100 2.5% and 4 *M* urea were set up and tested.

These agarose, urea and sorbitol (AUS) gels allowed the ready separation of *Acinetobacter* ß-lactamases by IEF.

Four ß-lactamases were distinguished in the original eight strains by AUS IEF; these were designated *Acinetobacter* chromosomal enzymes (ACE) 1–4. Of the five strains which produced ACE-1, four were *A. baumannii*, the most prevalent *Acinetobacter* species in clinical specimens (Bouvet and Grimont, 1987). The ACE-1 ß-lactamase has two pI bands of almost equal intensity at 8.6 and 8.8. Lesser satellite bands are seen sometimes at pI 9.8 and around 10.2, possibly resulting from carbamylation. One strain of *A. baumannii* possessed the ACE-2 ß-lactamase with a distinctive single band of pI 8.7. The ACE-3 ß-lactamase originated from an *A. junii* strain and has two bands of almost equal intensity with pIs of 8.1 and 8.2. Finally, ACE-4 came from an *A. lwoffii* strain, and also has two bands on AUS IEF with pIs of 7.3 and 7.7.

The above technique readily demonstrates the difference between the four chromosomal ß-lactamases, but is time consuming and very expensive. The results that it gives can be confirmed by separation of the ß-lactamases with ion-exchange chromatography. The basis of separation is by overall charge and is thus similar to IEF. Fast-Protein Liquid Chromatography (FPLC) allows rapid separation with the cation exchange column MONO S and the anion exchange column MONO Q. The fractions are tested for ß-lactamase activity with nitrocephin. FPLC separation showed clear differences between ß-lactamases ACE-1, ACE-3 and ACE-4, and suggested that there may be some relation between ACE-1 and ACE-2. Final confirmation of the distinct nature of these enzymes came from separation of the FPLC/ion exchange ß-lactamase peaks on a PAGE minigel. This gel system was devised to resolve basic proteins on polyacrylamide by Olsson and Tooke (1988), and was adapted to include a Phastgel Homogeneous 20 (SDS-free), run with reverse polarity, with visualisation of the ß-lactamases by nitrocephin (Hood and Amyes, 1991).

Now that *Acinetobacter* species were known to contain a series of chromosomal ß-lactamases, their properties could be determined. Morohoshi and Saito (1977) described an *Acinetobacter* ß-lactamase with an M_r of 30,000 by gel filtration, whereas Hikida et al. (1989) described an *Acinetobacter* enzyme with an M_r of 38,000 determined by SDS-PAGE after purification by ion exchange and gel filtration. The M_r values of the four ACE ß-lactamases described by Hood and Amyes (1991) were determined by gel filtration. The smaller enzymes were separated on Sephadex® G-75, and the larger enzymes were further separated on Sephacryl® S-300. ACE-3 was the smallest of the enzymes (M_r of 35,000). This is very similar to the values of both Morohoshi and Saito (1977) and Hikida et al. (1989). The ACE-2 ß-lactamase had an M_r of 60,500. The other two enzymes (ACE-1 and ACE-4) were both very large, always with an M_r of >500,000, and sometimes appearing >1,000,000. The size of these enzymes suggests that they are either multimeric or appear large because they are linked to other cell components which the purification techniques were incapable of removing, thereby offering a possible explanation for the difficulties encountered in solubilising these proteins. Hood and Amyes (1991) concluded by speculating that the size of these enzymes might reveal an intermediate step between penicillin binding proteins and true ß-lactamases.

Strains shown to possess just one of the ACE ß-lactamases were studied further by Hood and Amyes (1991). Their relative specific activities were determined against a number of substrates from partially purified cell-free extracts. All four enzymes were identified as cephalosporinases, although some possessed a little activity against penicillins. None of the enzymes had detectable hydrolysing activity against aztreonam or the third-generation cephalosporins ceftazidime or cefotaxime. This was particularly interesting as these *Acinetobacter* strains had been studied originally because they were aztreonam resistant (Table 8.1). Indeed all the host strains showed MICs of these cephalosporins that were higher than the recommended breakpoints. All the enzymes showed their maximum activity against cephaloridine and, except for ACE-4, good activity against cephradine. ACE-1 ß-lactamase showed the broadest spectrum of activity with some hydrolysis of cefuroxime (Table 8.2).

The affinities (K_m) of each of the ACE ß-lactamases for the substrate that all hydrolysed most readily, cephaloridine, were very similar, and were all within one order of magnitude (range 150–710 μM) (Table 8.3). Similarly the K_m of nitrocephin for the four enzymes varied within the same range of 150–280 μM (Table 8.3). These values are very similar to the K_m values published by both Morohoshi and Saito (1977) and Hikida et al. (1989). This close similarity showed that K_m values could not be used to distinguish the enzymes convincingly .

The designation of the original *Acinetobacter* cephalosporins as Bush class 1 (Bush, 1989) was based on their susceptibility to inhibitors. All four ACE ß-lactamases described by Hood and Amyes (1991) were highly susceptible to inhibition by cloxacillin, with all inhibited by 50% (ID_{50}) with <0.2 μM cloxacillin. Each was highly resistant to clavulanic acid, with ID_{50} values of >100 μM. There was little difference in the response of the enzymes to these inhibitors. However, this was not the case when aztreonam was used. Although there was no detectable hydrolysis of this ß-lactam by any of the enzymes, ACE-1, ACE-2 and ACE-3 were each inhibited by it. ACE-3 was the most sensitive, but both ACE-1 and ACE-2 were inhibited by <30 μM. ACE-4 was virtually insusceptible to the presence of aztreonam. All the original host strains had been aztreonam-resistant and, although their ß-lactamases did not

TABLE 8.2

Relative Rates of Hydrolysis of the ACE Chromosomal ß-Lactamases

	ACE chromosomal ß-lactamase			
Substrate	**ACE-1**	**ACE-2**	**ACE-3**	**ACE-4**
Nitrocephin	100	100	100	100
Cephaloridine	7.9	3.6	1.6	1.8
Cephradine	1.3	0.7	0.3	—
Cefuroxime	—	—	—	—
Cefotaxime	—	—	—	—
Ceftazidime	—	—	—	—
Aztreonam	—	—	—	—
Penicillin G	0.1	0.2	0.6	—
Ampicillin	—	—	—	—
Carbenicillin	—	—	—	—

Note: — = no detectable hydrolysis.

TABLE 8.3

Kinetic Values for the ACE Chromosomal ß-Lactamases (Data taken from Hood, 1991)

	Nitrocephin			Cephaloridine		
ß-Lactamase	V_{max}[a]	Km[b]	V_{max}/Km[c]	V_{max}[a]	Km[b]	V_{max}/Km[d]
ACE-1	0.83	220	100	0.005	150	5.8
ACE-2	0.20	250	100	0.01	260	5.0
ACE-3	0.11	150	100	0.0083	380	2.7
ACE-4	0.027	220	100	0.005	710	5.8

[a] µmol substrate hydrolysed/min/ml of ß-lactamase.

[b] µ*M*.

[c] Efficiency of hydrolysis value taken as 100%.

[d] Efficiency of hydrolysis for cephaloridine taken as a percentage of the value for nitrocephin.

hydrolyse this monobactam, three of them were able to bind it, which may partially account for their resistance.

All the ß-lactamases of *Acinetobacter* spp. were considered to be inducible by Sykes and Matthew (1976), Medeiros (1984) and Neu (1986). Morohoshi and Saito (1977) noted that the specific activities of the ß-lactamases in five strains of *Acinetobacter* "increased by about five- to tenfold" after treatment with 100 µg of benzylpenicillin, cephaloridine or 6-aminopenicillanic acid for 1 h at 37°C. Hikida et al. (1989) quoted these results and stated that the ß-lactamases of *Acinetobacter* spp. were inducible. Bauernfeind (1986) classified them as inducible or constitutive, whereas Sykes and Smith (1979) could find no evidence for induction, leading Joly-Guillou et al. (1988) to state that the case for inducibility has not been proven.

Extensive induction studies on the strains that produce the enzymes ACE-1 to ACE-4 by challenging them with cefoxitin at 0.25 × MIC by the method of Minami et al. (1980), failed to reveal any induction (Hood and Amyes, 1991). Furthermore, these authors stated that discussions with other workers had suggested that induction did not take place with any *Acinetobacter* strain when challenged with the 'classic' ß-lactamase inhibitors cefoxitin, imipenem or penicillin. However, Hood (1991) suggested that some *Acinetobacter* ß-lactamases may actually be inducible. Blechschmidt et al. (1989) and Borneleit and Kleber (1991) studied two strains of *Acinetobacter* (strains 69V and CCM5593) from the National Collection of Microorganisms at Brno in the former state of Czechoslovakia. It was not clear whether these strains were from environmental samples or clinical specimens, but it was suggested that the ß-lactamases from these strains were (1) inducible, and (2) predominantly extracellular. The extracellular location of these enzymes did not appear to result from leakage from the periplasm, as other periplasmic enzymes, such as alkaline phosphatase and glucose dehydrogenase, were not found extracellularly. Similarly, the presence of ß-lactamase in the surrounding medium did not appear to result from cell lysis because the cytoplasmic enzyme malate dehydrogenase could not be detected. The results showed an increase of ß-lactamase activity in the supernatant of crude cell-free extracts, which was accompanied by an increase in extracellular ß-lactamase and increased release of lipopolysaccharide. It was concluded that ß-lactamase export was dependent on over-production of the enzyme, coupled with a self-promoted perturbation of the outer membrane leading to a semiselective increase in membrane permeability. The case for inducibility remains unproven.

The contribution of these chromosomal ß-lactamases appears to be important in the expression of ß-lactam resistance, but probably works in concert with a permeability reduction and altered penicillin binding proteins which may already confer some inherent resistance. The acquisition of plasmid-encoded penicillinases does not seem to have been of paramount importance in the long-term ß-lactam resistance of this genus. As described earlier, there has so far been only one suggestion (Joly-Guillou and Bergogne-Bérézin, 1990) of a possible plasmid-encoded extended spectrum ß-lactamase in *Acinetobacter* spp., despite the fact that the progression to multiresistance in nosocomial *Acinetobacter* isolates in the last few years has been countered with a series of third-generation cephalosporins. This is perhaps surprising as the progenitors of many of these enzymes, the TEM-1 and TEM-2 ß-lactamases, have both been found in *Acinetobacter* strains.

8.2.3 Resistance to Carbapenems

A strain of *A. baumannii*, strain 6B92, isolated from a blood culture at the Royal Infirmary, Edinburgh in 1985, exhibited unusual resistance to imipenem, penicillin and cephaloridine, but had intermediate sensitivity to the third-generation cephalosporins ceftazidime and cefotaxime (Paton et al., 1993). Its imipenem MIC of 16 mg/L was particularly surprising because neither imipenem nor any other carbapenem had ever been used in this population; indeed imipenem was not introduced until some years later. Two ß-lactamases were produced by this strain: one with considerable activity and a sharp focusing band at pI 6.65; and a second, weaker ß-lactamase that focused poorly at a higher pI and was reminiscent of the cephalosporinases described above. The only other ß-lactamase found in *Acinetobacter* with a similar, but not identical, pI was CARB-5 (Joly-Guillou et al., 1988).

Both enzymes were partially purified and their sizes determined by gel filtration. The pI 6.65 enzyme had a molecular size of 23 kDa, while the high pI enzyme was 58 kDa, i.e., very similar to the ACE-2 cephalosporinase described by Hood and Amyes (1991). Enzyme kinetic data showed that the pI 6.65 enzyme had high affinity for penicillins and cephaloridine, but hydrolysis of these substrates was slow, particularly cephaloridine which was hydrolysed at only about a third of the rate for the penicillins. No hydrolysis was demonstrated by conventional spectrophotometric assays when imipenem, cefuroxime, cefotaxime and ceftazidime were used as substrates. However, although spectrophotometric assays are convenient for ß-lactamases, they may not always reveal hydrolysis. This may be particularly pertinent if hydrolysis is very slow or the substrate is unstable. A modification of the microbiological assay of Masuda et al. (1976) was developed by Paton et al. (1993) to provide a more sensitive technique for determining hydrolytic activity. This technique involves seeding an Isosensitest agar plate with *Staphylococcus aureus* strain NCTC 6571. Inhibitory zones for the antibiotics being tested are determined and the radii measured. Filter disks containing 15 µl of the undiluted or diluted ß-lactamase preparation are then applied to seeded plates at the expected periphery of the zone of sensitivity. Inactivation of the antibiotic is shown by growth of the indicator strain within the expected zone of inhibition (Figure 8.1). This technique clearly demonstrated hydrolysis and destruction of both imipenem and azlocillin by the pI 6.65 ß-lactamase. Cefuroxime, ceftazidime and cefotaxime were not inactivated, and this enzyme was therefore designated ARI-1 (*Acinetobacter* enzyme Resistant to Imipenem).

When nitrocephin was used as the substrate, the ARI-1 ß-lactamase was not inhibited readily by clavulanic acid, with an ID_{50} (Table 8.4)

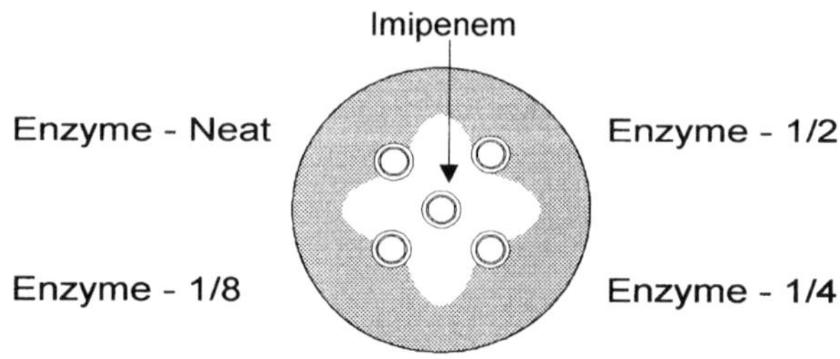

FIGURE 8.1
Microbiological assay of the ARI-1 ß-lactamase.

TABLE 8.4
Inhibitor Profile of the ARI-1 ß-Lactamase

Inhibitor	ID_{50} (μM)[a] ARI-1	ARI-2
*p*CMB	>100	>100
Clavulanic acid	100	0.5
EDTA	>10000	>10000
Imipenem	0.0875	>100

[a] Dose required to inhibit the ß-lactamase activity by 50%.

that was about 200-fold greater than that for TEM-1 (Paton et al. 1993). ARI-1 was inhibited neither by *p*CMB nor by high concentrations of EDTA, suggesting that metal ions play no role in the catalytic activity of this enzyme, and thus it is not a metallo-ß-lactamase. Furthermore, when imipenem or meropenem were used as inhibitors rather than substrates, both inhibited the action of ARI-1, but were without effect on the TEM-1 ß-lactamase, demonstrating that the carbapenems were bound much more strongly by ARI-1.

The inhibitor BRL42715 was a powerful inhibitor of ARI-1, with an ID_{50} of 1 n*M*. This inhibitor binds to most serine-based active site ß-lactamases, unlike clavulanic acid which binds preferentially to the class A enzymes. These results, coupled with the resistance to EDTA, suggest that ARI-1 is an unusual carbapenemase in that it is not a metallo-ß-lactamase, but rather a serine-based enzyme. Although the inhibitor results with clavulanic acid are inconclusive, and it is possible that ARI-1 is a class A ß-lactamase (Ambler, 1980), the results tend to suggest that it may be a class C enzyme. Sequencing of the ß-lactamase gene will confirm this, but whichever class of serine-based ß-lactamase ARI-1 belongs to, it is a very sinister development and augurs badly for the future of the carbapenems. This is the first, and remains the only report of an *Acinetobacter* ß-lactamase capable of hydrolysing and conferring resistance to carbapenems.

Conjugation studies in *Acinetobacter* species are not easy (Goldstein et al., 1983; Towner, 1991), but have failed uniformly to demonstrate any transfer or mobilisation of the ARI-1 ß-lactamase gene directly to *E. coli* or *Pseudomonas aeruginosa*. It could not be mobilised with the promiscuous *inc*P plasmid RP4, which readily entered *Acinetobacter* strains, nor could the isolated plasmid DNA be transformed into *E. coli* strain C600. However, we have recently demonstrated the direct conjugative transfer of the ARI-1 gene from its original *A. baumannii* host to an *A. junii* recipient (Scaife et al., 1995), and the same plasmid can be visualised in the donor and recipient strains. These latter observations suggest strongly that ARI-1 is a plasmid-encoded ß-lactamase. Further support for this view comes from some previous experiments, in which *A. baumannii* strain 6B92 was subcultured in increasing concentrations of ethidium bromide (Paton et al., 1993). The highest concentration in which the strain was able to grow was 64 mg/L. Examination of this culture revealed that 10% of cells had lost ARI-1 ß-lactamase activity as no band was visible by IEF, no imipenem-hydrolysing activity remained in the microbiological assay, the MIC of imipenem dropped by a factor of 100, and that of penicillins by a factor of 8. The MICs of the cephalosporins remained constant, suggesting that cephalosporin resistance derived from the chromosomal ß-lactamase that could still be visualised by IEF. These results describe the first plasmid-encoded imipenem (carbapenem) resistance in *Acinetobacter* spp., and the first plasmid-encoded imipenem resistance found in any bacterial species isolated in Europe.

A summary of all the ß-lactamases described in *Acinetobacter* spp. is shown in Table 8.5.

8.2.4 Other Mechanisms of ß-Lactam Resistance

The multiresistant nature of *Acinetobacter* spp. suggests that impermeability is likely to be an important factor. Sato and Nakae (1991) examined the outer membrane proteins of wild-type strains of *Acinetobacter* that had not yet acquired clinically important levels of ß-lactam resistance. Measurement of the cephalosporinase activity of intact cells and cell-free extracts showed that ß-lactam hydrolysis by intact *Acinetobacter* cells was about 0.02% that of the cell-free extracts. When compared with results obtained with *Pseudomonas* strains, which are well known to confer resistance by restricting the passage of antibiotics through their outer membrane porins (Yoshimura and Nikaido, 1982; Parr et al., 1987), it was concluded that the outer membrane of *Acinetobacter* was even less permeable than that of *P. aeruginosa*.

TABLE 8.5

Summary of ß-Lactamases Described in *Acinetobacter* spp.

Enzyme or strain	Location of gene	Predominant substrate	Molecular size (kDa)	Ref.
TEM-1	Plasmid	Penicillin	29	1
TEM-2	Plasmid	Penicillin	29	2
CARB-5	Plasmid	Penicillin	29	3
ARI-1	Plasmid	Carbapenem	23	4
NCTC7844	Chromosome	Cephalosporin	30	5
ML4961	Chromosome	Cephalosporin	38	6
ACE-1	Chromosome	Cephalosporin	*ca.*500	7
ACE-2	Chromosome	Cephalosporin	60.5	7
ACE-3	Chromosome	Cephalosporin	32.5	7
ACE-4	Chromosome	Cephalosporin	>1000	7
SHV-like	Unknown	Penicillin	Unknown	8
SHV-like	Unknown	Penicillin	Unknown	9

Note: References: (1) Philippon et al. (1980); (2) Devaud et al. (1982); (3) Bauernfeind (1986); (4) Paton et al. (1993); (5) Morohoshi and Saito (1977); (6) Hikida et al. (1989); (7) Hood and Amyes (1991); (8) Joly-Guillou and Bergogne-Bérézin (1990); (9) Vila et al. (1993).

Sato and Nakae (1991) also examined the penetration of zwitterionic ß-lactam antibiotics into liposomes created from *Acinetobacter* outer membranes, and found that the carbapenems imipenem and meropenem diffused at about 10% the rate of glucose. In a similar experiment with *E. coli*, the rate was >50%. When the results were quantified in relation to unit protein concentration, it was demonstrated that *Acinetobacter* allowed diffusion of the carbapenems at about 1–2% of the efficiency of *E. coli* B. When penetration of cephalosporins was measured, diffusion into *A. calcoacaeticus* at the even slower rate of 2–5% of the rate of glucose was observed. It was hypothesised that the remarkably low permeability of *Acinetobacter* might result from either small outer membrane pore size, limited porin production, or a combination of both. When the outer membrane proteins of *Acinetobacter* were examined by separation on an SDS-PAGE gel, it was found that the porins comprised <5% of the total outer membrane protein; this compares with the porins comprising 60% of the outer membrane protein in *E. coli* B (Rosenbusch, 1974). In addition, the demonstration of much slower diffusion of carbapenems and some sugars allowed the conclusion that the porins, besides being less numerous, were also of a particularly small size. Thus the inherent insusceptibility of *Acinetobacter* spp. to ß-lactams may result from a combination of small outer membrane pore size and limited porin production. Sato and Nakae (1991) also claimed that this was probably the first discovery of a naturally occurring wild-type bacterium that produced a minimal amount of porin.

It was uncertain whether a smaller number of small-sized porins from a wild-type strain could be modified further to confer increased levels of ß-lactam resistance. Obara and Nakae (1991) examined two clinical strains of *Acinetobacter*, strains 4751 and 4760. Strain 4751 was a ß-lactam resistant isolate, whereas strain 4760 was generally ß-lactam-sensitive. Three cephalosporin-resistant mutants of strain 4760, TNA001, TNA002 and TNA003, were isolated by selection on high concentrations of cefoxitin, cefoperazone or ceftazidime, respectively. Examination of cell-free extracts revealed that the resistant clinical strain 4751 produced 77.4-fold as much ß-lactamase as the sensitive clinical strain 4760 (4.8 μmol of cephalothin hydrolysed/min/mg protein compared with 62 nmol). These ß-lactamase specific activities were unaffected if the strains were pre-induced with cephaloridine, imipenem or cefoperazone. In contrast, the resistant mutants of strain 4760 did not produce significantly elevated levels of ß-lactamase, as none of the three mutants had a specific activity of >94 nmol cephalothin hydrolysed/min/mg protein, and they could not be induced. Increased ß-lactamase production was therefore not responsible for the cephalosporin resistance in the laboratory mutants, whereas this was a major factor in the resistant clinical strain.

Examination of the penicillin binding proteins (PBPs) of *Acinetobacter* has revealed that they are unlike those in *E. coli* and *P. aeruginosa*, and are most similar to those of the genus *Bacteroides* (Piddock and Wise, 1986). Obara and Nakae (1991) found six PBPs in *Acinetobacter*, which were designated PBPs 1a, 1b, 1c, 2, 3, 4, and were 94, 92, 86, 74, 59 and 42 kDa in size, respectively. All three of the mutant strains TNA001, TNA002 and TNA003 had a diminished amount of PBP 1a, and the mutant selected on ceftazidime, TNA003, also showed 60% elevated production of PBP 1b. The amount of cefoxitin required for 50% saturation of PBP 3 from the mutant selected on cefoxitin, TNA001, was 30-fold higher than for the sensitive parent strain 4760, suggesting that PBP 3 was the major target for cefoxitin. For the cefoperazone-resistant mutant, TNA002, the cefoperazone concentrations required for 50% saturation of PBPs 1a, 1b, 2 and 3 were >4.4, 4.5, 3.3 and 3.3 times higher, respectively, than those for strain 4760. TNA003 showed little change for any PBP when 50% saturation assays were performed with ceftazidime.

Obara and Nakae (1991) used competition assays with cefoxitin, cefoperazone and ceftazidime to establish the affinity of the cephalosporins for the PBPs in the mutants. The amounts of

[^{14}C]-benzylpenicillin bound to the TNA001 PBP 3, the TNA002 PBP 1b and the TNA003 PBP 1b were higher than those bound to the corresponding PBPs of the parent strain 4760, while a lower level of cefoxitin bound to the TNA001 PBP 3, a lower level of cefoperazone bound to the TNA002 PBP 1b, and a similarly reduced amount of ceftazidime bound to the TNA003 PBP lb. Obara and Nakae (1991) also examined the outer membrane permeability of these *Acinetobacter* strains. When the diffusion rate of *p*-nitrophenyl phosphate (*p*NPP) through reconstituted liposome membranes was measured, it was found that the amount of *p*NPP that diffused through the membranes of mutants TNA001, TNA002 and TNA003 was 65, 43 and 78%, respectively, of that for strain 4760, demonstrating that these mutants retarded the passage of *p*NPP compared with the sensitive parent strain. This was coupled with a decrease in the amount of the 46.5-kDa outer membrane protein.

The conclusion of this work was that *Acinetobacter* strains express ß-lactamases that are produced constitutively and cannot be induced. However, in resistant clinical isolates, the level of ß-lactamase expression can increase markedly. Bearing in mind that *Acinetobacter* spp. can produce several ß-lactamases, it is sometimes difficult to determine which ß-lactamase has increased in production. The response and production of individual ß-lactamases, and the role of their individual specific activities in the resistance found in clinical isolates, has yet to be determined. It is clear from Obara and Nakae (1991) that, whatever the cause of increased ß-lactamase production, it is not a simple response and probably has taken a number of mutational steps to develop. The initial stages in the development of resistance appear to potentiate ß-lactamase activity without increasing production above the basal level in the sensitive strain. This may be achieved in one of two ways, either by restricting the entry of ß-lactam into the cell or by making the PBP less susceptible to binding, thus providing more opportunity for the available ß-lactamase to destroy the antibiotic. The PBPs underwent some marked changes in the three mutants studied. In one case, PBP production was enhanced so that more antibiotic was required to bind before inhibition of the cell could occur, and in the remainder the PBPs were altered so that they were less susceptible to binding per unit molecule, which essentially gives the same effect. Thus a combination of less susceptible PBPs and available ß-lactamase appears responsible for resistance. The contribution of the porin numbers in the outer membrane is less easy to define, but it does seem to affect the predisposition of this genus to develop resistance. In this study, the numbers of porins were diminished in these mutants, but not to a great extent, especially bearing in mind the small number of porins in *Acinetobacter* in the first place.

8.3 AMINOGLYCOSIDE RESISTANCE

8.3.1 Aminoglycoside-Modifying Enzymes

Aminoglycoside resistance in Gram-negative bacteria results most commonly from the action of plasmid- or transposon-encoded aminoglycoside-modifying enzymes which render aminoglycosides inactive. Three different types of modifying enzymes have been described: acetyltransferases (AAC), adenyl- or nucleotidyltransferases (AAD or ANT), and phosphotransferases (APH). Enzymes within each group are classified according to the aminoglycoside site of modification. Thus, there are four classes of acetyltransferases, AAC(1), AAC(3), AAC(6′) and AAC(2′), five of adenyltransferases, ANT(2″), ANT(3″), ANT(4′), ANT(6) and ANT(9), and five of phosphotransferases, APH(2″), APH(3′), APH(3″), APH(6) and APH(4). Several of the enzymes can be subdivided further according to their aminoglycoside resistance profile (designated by a Roman numeral suffix) and unique protein nature (designated by a lower-case alphabetical suffix). Hence, AAC(6′)-Ia and AAC(6′)-Ib express identical aminoglycoside resistance profiles, but differ in their protein sequence (Shaw et al., 1993).

The identification of specific aminoglycoside-modifying enzymes can, in many instances, be achieved by recognition of a specific aminoglycoside resistance profile (AGRP) (van der Klundert et al., 1984). However, interpretation of data may be complicated by the presence of several different enzymes within a single isolate, so that confirmation of the presence of a particular enzyme may require extraction and purification, followed by substrate profile analysis of each individual purified enzyme. More recently, DNA probes specific for the structural genes encoding a number of different aminoglycoside-modifying enzymes have been developed (Tenover et al., 1984; 1989; Young et al., 1985; Barg, 1988; Lovering et al., 1988; Lambert et al., 1990) and used successfully to characterise the dissemination of specific aminoglycoside resistance genes in hospital epidemics (Tenover et al., 1984; O'Brien et al., 1985; Barg, 1988; Lovering et al., 1988; Chaslus-Dancla et al., 1989; Licitra et al., 1989; Lambert et al., 1990).

8.3.2 Prevalence of Aminoglycoside-Modifying Enzymes

Aminoglycosides are used widely for the treatment of *Acinetobacter* infections, and increasing numbers of highly resistant strains have been reported since the late 1970s. Several groups of workers have investigated resistant strains for the presence of modifying enzymes, and all three types of transferases have been identified within clinical strains of *Acinetobacter* (Table 8.6). The distribution of aminoglycoside-modifying enzymes within clinical isolates of *Acinetobacter* spp., as reported by Shaw et al. (1991; 1993), is summarised in Table 8.7.

In the first survey, 142 isolates collected between 1985 and 1988 from hospitals in the U.S., Belgium and Argentina were investigated and the incidence of particular AGRPs was compared to hybridisation results obtained with specific gene probes (Shaw et al., 1991). Although a good correlation was found between the AGRP and the presence of a particular gene, the results highlighted the fact that a single enzyme, mediating resistance to a defined spectrum of aminoglycosides, can be encoded by a range of different genes, many of which remain uncharacterised and therefore lack specific DNA probes to enable their detection. Results from this survey also suggested geographic variations in the incidence of particular genes, e.g., the AAC(3)-Ia gene was found frequently in *Acinetobacter* strains from Belgium (36 of 45 strains), but was observed less frequently in strains from the U.S. (3 of 17 strains), and not at all in the strains from Argentina. Furthermore, the hybridisation data confirmed that some strains contained more than one aminoglycoside resistance gene, with as many as six different resistance genes being identified in some isolates.

The second survey investigated 206 aminoglycoside-resistant *Acinetobacter* isolates collected between 1987 and 1991 in the U.S. (Shaw et al., 1993). Six different modifying enzymes were identified according to the AGRP expressed. Acetyltransferases modifying the 3-amino group, AAC(3)-I and AAC(3)-II, were identified in 29.6 and 21.3% of isolates, respectively. Phosphotransferases modifying the 3'-amino group, APH(3')-I and APH(3')-VI, were identified in 49.5 and 35.4% of isolates, respectively. Interestingly, the adenylating enzyme ANT(2")-I, which had been identified only rarely in the earlier study, was found to be quite prevalent amongst these later isolates, with 15.5% of isolates expressing an AGRP indicative of the presence of the enzyme (Shaw et al., 1993).

More recently, a novel gene termed *aac(6')-Ig,* identified in *Acinetobacter haemolyticus* and responsible for amikacin resistance, has been characterised (Lambert et al., 1993). A DNA probe specific for the gene was developed and used to screen 162 clinical strains of *Acinetobacter* resistant to amikacin. Interestingly, only *A. haemolyticus* strains hybridised to the probe, and no hybridisation was observed with other strains of *Acinetobacter*, including *A. baumannii, A. lwoffii, A. junii* and *A. johnsonii.* Lambert and colleagues suggested that the *aac(6')-Ig* gene is species-specific and proposed that its presence may be used to facilitate identification of *A. haemolyticus.*

TABLE 8.6

Aminoglycoside-Modifying Enzymes Identified in *Acinetobacter* spp.

Enzyme	Resistance profile (AGRP)	Ref.
Acetylating:		
AAC(6′)	Tob Dbk Ntl 6′Ntl 2′Ntl Siso Amk	Shannon et al. (1978) Murray and Moellering (1979) Lambert et al. (1990; 1993)
AAC(2′)I	Tob Dbk Ntl 6′Ntl Gm	Dowding (1979)
AAC(3)I	Apr Lvdm Prm Rsm (But) (Neo)	Bergogne-Bérézin et al. (1980) Vila et al. (1993)
AAC(3)II	Tob Dbk Ntl 6′Ntl 2′Ntl Siso Gm	Murray and Moellering (1980)
[AAC(3)V]		Elisha and Steyn (1991a) Shaw et al. (1993)
AAC(3)IV	Tob Dbk Ntl 6′Ntl 2′Ntl Siso Gm Apr	Shaw et al. (1993)
Adenylating:		
ANT(3″)I	Sm Spcm	Shannon et al. (1978)
[AAD(3″)(9)]		Murray and Moellering (1980) Devaud et al. (1982) Goldstein et al. (1993) Vila et al. (1993)
ANT(2″)I	Tob Dbk Siso Gm Km	Murray and Moellering (1980)
[AAD(2″)]		Elisha and Steyn (1991a) Shaw et al. (1993)
Phosphorylating:		
APH(3′)I	Km Neo Prm Rsm Lvdm GmB	Bergogne-Bérézin et al. (1980) Devaud et al. (1982) Goldstein et al. (1983) Shaw et al. (1993)
APH(3′)II	Km Neo Prm Rsm But GmB (Amk)	Murray and Moellering (1979)
APH(3′)III	Km Neo Prm Rsm Lvdm But GmB Amk Isp	Murray and Moellering (1979) Joly-Guillou et al. (1987)
APH(3′)VI	Km Neo Prm Rsm But GmB Amk Isp	Martin et al. (1978) Lambert et al. (1988; 1990) Vila et al. (1993) Shaw et al. (1993)
APH(3″)I	Sm	Elisha and Steyn (1989)

Note: Amk, amikacin; Apr, apramycin; But, butirosin; Dbk, dibekacin; Gm, gentamicin; GmB, gentamicin B; Isp, isepamicin; Km, kanamycin; Lvdm, lividomycin; Neo, neomycin; Ntl, netilmicin; 2′Ntl, 2′-*N*-ethylnetilmicin; 6′Ntl, 6′-*N*-ethylnetilmicin; Sm, streptomycin; Prm, paromycin; Rsm, ribostamycin; Siso, sisomicin; Spcm, spectinomycin; Tob, tobramycin. Parentheses indicate enzymic activity detectable only *in vitro*.

TABLE 8.7

Prevalence of Aminoglycoside-Modifying Enzymes in Clinical Isolates of *Acinetobacter* spp.

Enzyme [DNA probe]	% AGRP [probe] positive	
	Shaw et al. (1991)	Shaw et al. (1993)
AAC(3)I	47.2%	29.6%
[AAC(3)Ia]	[27.5%]	
AAC(3)II / AAC(3)V	14.1%	21.3%
[AAC(3)Va]	[5.6%]	
AAC(3)IV		0.5%
AAC(6′)I	26.8%	
[AAC(6′)Ia]	[4.9%]	
[AAC(6′)Ib]	[0.0%]	
ANT(2″)I	1.4%	15.5%
[ANT(2″)Ia]	[1.4%]	
[ANT(2″)Ib]	[0.0%]	
APH(3′)I	55.6%	49.5%
[APH(3′)Ia]	[29.6%]	
[APH(3′)Ib]	[0.0%]	
APH(3′)VI		35.4%

8.3.3 Plasmid-Borne Aminoglycoside Resistance Genes

Few studies have investigated the genetic nature of aminoglycoside resistance in *Acinetobacter* spp. successfully. This is a result, primarily, of the lack of an efficient plasmid transfer model for *Acinetobacter* spp. which, although apparently able to acquire plasmids from Enterobacteriaceae readily, are less able to transfer indigenous plasmids to *E. coli*, the standard recipient in most laboratory matings (Towner, 1991). Murray and Moellering (1980) were the first to suggest that aminoglycoside-modifying enzymes identified in *Acinetobacter* spp. may be plasmid-borne, and reported that treatment of a multiply resistant *Acinetobacter* (strain H-S) with novobiocin resulted in the loss of several resistance genes, including those for kanamycin and gentamicin resistance. The cured strain (H-SC) no longer expressed the aminoglycoside-modifying enzymes ANT(2") and APH(3")-III, and physical loss of a large plasmid from the cured, sensitive strain was demonstrated by agarose gel electrophoresis. Loss of genes encoding the AAC(3) and APH(3')-III modifying enzymes was shown to occur in a second cured isolate (H-DC1), but this was not associated with detectable loss of plasmid DNA. It was concluded from this evidence that the genes for aminoglycoside resistance in the two strains were most likely to be plasmid-borne, though no transfer of resistance was demonstrated, and a possible chromosomal location could not be excluded.

Early studies confirming a plasmid location for aminoglycoside resistance genes in *Acinetobacter* reported extremely low plasmid transfer frequencies to *E. coli* recipients (Gomez-Lus et al., 1980; Goldstein et al., 1983). Gomez-Lus et al. (1980) demonstrated a plasmid location for *aac(3)*, responsible for gentamicin resistance, and *aac(6')*, responsible for amikacin and tobramycin resistance, in *Acinetobacter* clinical isolates. Goldstein et al. (1983) reported a 167-kb, *inc*C group plasmid, pIP1031, that encoded streptomycin and spectinomycin resistance mediated by an AAD(3")(9) enzyme, i.e., ANT(3"), and kanamycin resistance mediated by an APH(3')-I enzyme.

In a survey of 115 amikacin-resistant clinical isolates of *Acinetobacter* spp. collected from hospitals in France, Israel and Tunisia, Lambert et al. (1990) investigated the prevalence and genetic location of the *aphA6 (aph(3')-VIa)* resistance gene encoding the APH(3')-VI aminoglycoside-modifying enzyme. Dot-blot hybridisation experiments indicated that the *aph6A* gene was present in 95% of isolates, and transfer of amikacin resistance to *Acinetobacter* recipients was observed from 49 strains (43% of isolates). Plasmid restriction endonuclease fingerprints and hybridisation data indicated that several different plasmids were involved in the spread of the *aph6A* gene, and a possible transposon location of the gene was suggested.

Molecular genetic techniques were used by Elisha and Steyn (1991a) to identify plasmid-located aminoglycoside resistance genes. A gene encoding an AAC(3)-II enzyme, responsible for gentamicin and tobramycin resistance, was identified on a 50–60 kb plasmid present in a multiresistant clinical isolate of *Acinetobacter*, designated strain SAK. Sequence analysis of the gene's control region showed little similarity to the published AAC(3)-II gene control sequence (Vliegenthart et al., 1989) and it was suggested that the gene may contain a catabolite-sensitive promoter. It was speculated that the unusual occurrence of a promoter regulated by the availability of energy sources may relate to a possible soil origin of the SAK strain, since soil environments where antibiotics are produced have been suggested to impose special metabolic demands on microorganisms (Shaw, 1983). Strain SAK also hybridised with a DNA probe specific for the gene encoding the AAD(2") modifying enzyme. However, Northern blot analysis of RNA transcripts synthesised by strain SAK revealed only AAC(3)-II transcripts, and no hybridisation of RNA transcripts was observed with the AAD(2")-specific probe. RNA transcripts specifying the AAD(2") enzyme could not be detected, even after culture of the strain in increasing concentrations of kanamycin, suggesting that resistance to this aminoglycoside was not mediated by the AAD(2") enzyme and must therefore result from either a mutated AAC(3)-II gene product or a third aminoglycoside resistance gene within the strain. The lack of expression of the gene encoding the AAD(2") enzyme illustrates the potential problems arising from the use of DNA probes to study the epidemiology of aminoglycoside resistance genes, since a positive hybridisation result does not necessarily confirm the mechanism of resistance to a particular aminoglycoside in a resistant strain.

8.3.4 Transposon-Borne Aminoglycoside Resistance Genes

Transposon-encoded aminoglycoside resistance in *A. calcoaceticus* was first demonstrated by Devaud et al. (1982). A multiply resistant isolate of *Acinetobacter* (strain HK302), responsible for an outbreak of respiratory tract infections in patients in an intensive care unit in the late 1970s, was found to encode three aminoglycoside-modifying enzymes, AAC(3)-I, APH(3')-I and ANT(3"), responsible for gentamicin, kanamycin and streptomycin/spectinomycin resistance, respectively. Strain HK302 was also resistant to chloramphenicol, tetracycline and sulphamethoxazole. Transfer of resistance to *E. coli, P. aeruginosa* or *A. calcoaceticus* recipients could not be demonstrated by conjugation. However, mobilisation of gentamicin, streptomycin, chloramphenicol, sulphamethoxazole and, possibly, tetracycline resistance was achieved following introduction of the *inc*P group plasmid RP4 into a kanamycin-sensitive derivative of the original wild-type HK302 strain. Following mobilisation, plasmid RP4 was shown to have increased in size by 16 Mda, suggesting the acquisition of a multiple-resistance transposon. It was proposed that the transposon was chromosomally located in strain HK302, and that it had been imported originally on a plasmid, acquired from multiply resistant Gram-negative hospital flora, which was unstable in *Acinetobacter* strain HK302.

Hybridisation analysis of the multiple-resistance plasmid pIP1031, identified in a clinical isolate of *A. calcoaceticus,* strain BM2500 (Goldstein et al., 1983), revealed the presence of two deletion derivatives of the insertion sequence IS*15*, one of which was confirmed to be within the same *Eco*RI restriction fragment as the gene encoding an APH(3')-I enzyme. The occurrence of this APH(3')-I gene within other well-characterised compound transposons such as Tn*6* (Berg et al., 1975) and Tn*903* (Oka et al., 1981) resulted in the speculation that the APH(3')-I gene of pIP1031 may also be contained within a compound transposon (Goldstein et al., 1983).

8.4 4-QUINOLONE RESISTANCE

The fluorinated 4-quinolones must be regarded as one of the major contributions to the control of severe nosocomial infections caused by

Gram-negative bacteria. As described in the previous chapter, the emergence of *Acinetobacter* spp. as important hospital pathogens has occurred at the same time as increased reliance on 4-quinolones for the treatment of serious infection. It is worth considering whether these two facts are coincidental, or whether the use of one leads to the other.

The enigma of the development of 4-quinolone resistance is that it is often quite difficult to demonstrate in the laboratory, and this finding has often been extrapolated to suggest that resistance will be rare in the clinical situation. This is true for bacteria such as *E. coli*, but is not true for non-fermentative Gram-negative bacteria such as *P. aeruginosa* and *Acinetobacter* spp. The reasons for this may be twofold: first, the traditional method for determination of mutation frequencies, whereby a population of bacteria is plated on to a medium containing a concentration of antibacterial drug that is a multiple of the MIC, e.g., 5× or 10×, will only reveal any mutants that were pre-existing in the population. Extrapolation of data from this technique into the clinical situation assumes that a similar challenge takes place, i.e., with a fixed high concentration of antibacterial drug. This is, of course, not the case. All clinical antibacterial administration causes the concentration at the site of infection to rise and fall until the next dose of antibacterial is given. During the trough period, the level of antibacterial may often fall below the MIC. When a bacterial culture is treated with sub-MIC concentrations of an antibacterial, all cells are able to divide; however, the resistant strains are able to divide more readily, and often much more quickly, than the partially sensitive strains. Therefore, even when the original population contained no resistant mutants, if a mutation occurred early in the sub-MIC challenge, its progeny could soon outgrow the partially inhibited sensitive parent cells. This is very easy to demonstrate in the development of 4-quinolone resistance. Mutation frequencies should be measured against the number of cell divisions, not the number of cells present; sub-MIC challenge increases the number of cell divisions compared with super-MIC challenges. Second, non-fermenters often have reduced uptake mechanisms, and are thus already less susceptible to antibacterial challenge than other bacteria. This will allow them to survive and divide for a longer period than their competitors. Such a competitive edge will be revealed, in population dynamics, as the organism having a predisposition towards the development of resistance. This is the response observed with *Acinetobacter* spp. in the development of antibiotic resistance as a whole, and to the 4-quinolones in particular.

The fact that *Acinetobacter* spp. can develop 4-quinolone resistance readily is not disputed, and although the precise mechanism is virtually unknown, some proposed mechanisms have been suggested, largely on the basis of findings with other bacteria. DNA gyrase (topoisomerase

II), the target of 4-quinolone action, is composed of four subunit proteins; two identical α-subunits and two identical ß-subunits. The purpose of this enzyme is to introduce negative supercoils into the bacterial DNA. There is no closely equivalent mammalian enzyme, so any antibacterial capable of inhibiting this enzyme is likely to be selective. Bacterial resistance to 4-quinolones in other genera has often been attributed to changes in the structure of the DNA gyrase subunits, usually by *gyrA* mutations causing alterations in the α-subunit. In most species, these single or multiple changes in the α-subunit have conferred decreased susceptibility to all 4-quinolones by the same proportion. Thus any discrimination between the individual members of the fluorinated 4-quinolones does not occur at the drug-target interaction. Vila et al. (1994) used the polymerase chain reaction (PCR) to amplify DNA surrounding the active site region of the *gyrA* gene. This region was amplified from 13 clinical isolates of *A. baumannii*, with a range of ciprofloxacin MICs from 0.25–64 mg/L, and the resulting PCR fragments were sequenced. In the sensitive bacteria, 87 nucleotide differences, correlating with 13 amino acid differences, were found compared with the same 290-bp fragment from *E. coli*. The residues Gly-81, Ser-83, Ala-84 and Gln-106, whose substitution leads to 4-quinolone resistance in *E. coli*, were all conserved in the sensitive *A. baumannii* strains. All nine isolates of *A. baumannii* with ciprofloxacin MICs of >2 mg/L showed a substitution of Ser-83 to leucine and Ala-84 to proline. Four strains with an MIC of 1 mg/L did not show any change at Ser-83, but one exhibited a change from Gly-81 to valine. The results correlated with those from *E. coli* in that substitution of Ser-83 was responsible for, or at least contributed to, ciprofloxacin resistance in *A. baumannii*. *GyrB* mutations, causing changes in the ß-subunit, occur less frequently and rarely result in such high levels of resistance. None has been examined in *Acinetobacter* spp.

Experience with other Gram-negative bacterial species suggests that DNA gyrase mutations are not the first response to challenge with these drugs. Mutations in the DNA gyrase subunits confer resistance to 4-quinolones and a few related compounds, such as the coumarins, but not to unrelated compounds. Thus they may be of less assistance to the cell's survival than mutations that confer cross-resistance to other antibacterial drugs. As described previously, *Acinetobacter* strains are less permeable to antibacterials than many Gram-negative organisms, and 4-quinolone resistance can also be conferred by changes in the outer membrane that result in decreased uptake. Indeed, it has been shown in *E. coli* and *P. aeruginosa* that selection of resistance by 4-quinolones can result in cross-resistance to ß-lactams (Neu, 1988). This suggests strongly that resistance results from alterations in the outer membrane, leading to decreased uptake. Studies on the development of 4-quinolone resistance in *P. aeruginosa* suggest that changes in the outer membrane proteins are an early development in the build-up of resistance genes contributing to 4-quinolone resistance (Quibell et al., 1993), and this is also likely to be true for *Acinetobacter* spp. Muller-Serieys et al. (1989) showed that there has been an increase in the proportion of *Acinetobacter* isolates with combined resistance to all ß-lactams, all aminoglycosides and quinolones. It is not possible, without detailed mechanistic analysis, to attribute this development to reduced permeability, but it remains a strong contender.

Ng et al. (1994) cloned the *S. aureus norA* gene, which produced resistance to norfloxacin and other hydrophilic quinolones, and reduced norfloxacin accumulation in intact cells. The action of the gene was energy dependent, suggesting active drug efflux as the mechanism of resistance. The results suggested that norfloxacin efflux was mediated by a carrier with a particularly high affinity for norfloxacin, although ciprofloxacin and ofloxacin could inhibit an uptake assay competitively. It was further suggested that the NorA protein itself functions as a drug transporter that is coupled to the proton gradient across the cell membrane. How much of this proposed efflux system is relevant to *Acinetobacter* spp. is unknown, but a similar system should be considered as a potential resistance mechanism. One reason is that this may be the mechanism of 4-quinolone resistance that is most likely to succeed as a plasmid-mediated mechanism. Tetracycline resistance, encoded by plasmids, is invariably mediated by an efflux system, and there is no obvious genetic mechanism by which the *gyrA* gene could function from a clinical plasmid location. Indeed, Robillard (1990) and Piddock et al. (1992) demonstrated that the introduction of a sensitive plasmid-encoded *gyrA* gene into *P. aeruginosa* abolished all resistance conferred by a resistant chromosomal DNA gyrase. In other words, the introduced sensitive DNA gyrase was dominant over the resistant enzyme. It has always been presumed that the reverse would also be true, i.e., that the introduction of a plasmid-encoded altered DNA gyrase gene which was 4-quinolone resistant would be nullified by the presence of a sensitive chromosomal enzyme. This conviction may not be infallible, and Soussy et al. (1993) have questioned the whole validity of this plasmid-dominance test. There is not much experimental evidence, and plasmid-dominance has been examined only with fully sensitive DNA gyrase genes. If the chromosomal DNA gyrase gene itself is partially resistant to 4-quinolone binding, the effects appear to be unknown. The complete lack of plasmid-encoded nalidixic acid resistance in all bacterial species tends to suggest that plasmid-encoded 4-quinolone resistance will not emerge. However, nalidixic acid was only used rarely in many areas, and the selection pressure for plasmid-encoded resistance may have just been insufficient. In contrast, fluorinated 4-quinolones have been used widely wherever they have been introduced, and the selection pressure for a plasmid-encoded resistance mechanism, if it can exist, is very strong. If it can emerge, it will.

8.5 TRIMETHOPRIM RESISTANCE

8.5.1 Overview of Common Resistance Mechanisms

Trimethoprim acts by competitive inhibition of the bacterial enzyme dihydrofolate reductase (DHFR), which is required for the reduction of dihydrofolate to the essential cofactor tetrahydrofolate. High-level trimethoprim resistance (MIC of >1000 mg/L) in Gram-negative bacteria is mediated generally by the plasmid- or transposon-encoded synthesis of an additional insusceptible DHFR which enables the trimethoprim-sensitive host cell enzyme to be bypassed (Amyes and Towner, 1990). A range of plasmid-encoded DHFRs have been distinguished by their biochemical and molecular characteristics (Amyes et al., 1992), and sequence determination of the respective genes has enabled gene and oligonucleotide probes to be developed (e.g., Amyes and Towner, 1990; Young and Hillyear, 1994). These have been used to determine the prevalence of various trimethoprim resistance genes in clinical isolates (e.g., Heikkila et al., 1990; Towner et al., 1990; Young and Hillyear, 1994).

To date, the majority of surveys of Gram-negative bacterial pathogens have identified the type Ia DHFR as the predominant enzyme responsible for high-level trimethoprim resistance (Steen and Sköld, 1985). Initially, the successful spread of this gene was attributed to its presence within the highly promiscuous transposon Tn*7* (Barth et al., 1976), which is unusual in possessing two different transposition pathways that allow it to integrate at relatively low frequency into a wide range of different plasmids, and at a significantly higher frequency into a specific insertion site within the chromosome of a range of different bacterial hosts (Craig, 1989). More recently the *dhfrIa* gene has been identified within the ubiquitous family of Tn*21*-like transposons (Sundström and Sköld, 1990), thereby further promoting its dissemination within bacterial populations.

Several other *dhfr* genes have also been shown to reside within transposons. All those which have been characterised can be classified as either Tn*7*-like or Tn*21*-like in structure. The extended array of different *dhfr* genes which are found incorporated within Tn*7* and Tn*21*-like transposons results from the presence of integron structures within these two classes of transposon. Integrons can be defined as genetic elements, found within plasmids and transposons, which

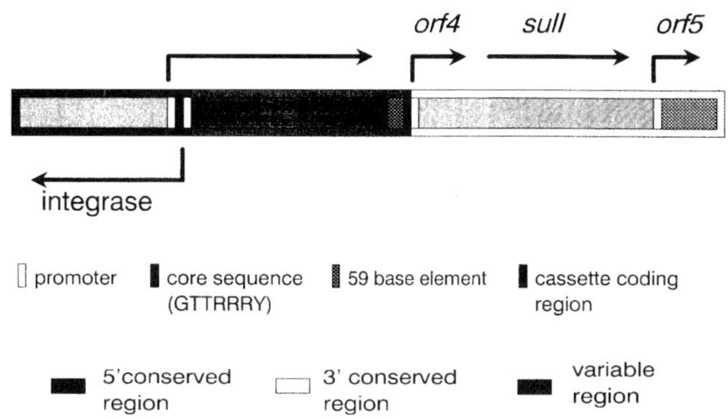

FIGURE 8.2
Basic Tn*21* integron structure.

encode a site-specific recombination system enabling the insertion, deletion and rearrangement of discrete genetic cassettes within the integron element (Stokes and Hall, 1989). The majority of gene cassettes identified within integron structures encode resistance to antibiotics (Bissonnette and Roy, 1992). The integron structure identified within Tn*21* and related transposons comprises a variable central region of DNA flanked by two conserved regions (Figure 8.2). The 3' conserved segment generally includes the sulphonamide resistance gene, *sulI*, flanked by two open reading frames *orf4* — identified as a *qac* gene encoding a transporter protein mediating resistance to antiseptics and disinfectants (Paulsen et al., 1993) — and *orf5* (function unknown). The 5' conserved segment encodes an integrase (*int*) that mediates the site-specific integration, deletion or rearrangement of genetic cassettes into a core site within the variable region of the integron (Collis and Hall, 1992; Collins et al., 1993). The 5' conserved segment also provides the promoter for transcription of integrated cassettes in the variable region. Each genetic cassette comprises three conserved features: a central coding region, all except the last seven bases of a 59-bp element (comprising a short imperfect repeat sequence associated with the 3' end of the gene), and a core site comprising the last seven bases of the 59-bp element located 5' to the gene. These 59-bp elements act as recombination hot-spots (RHS) for integrase action (Hall et al., 1991). The integron identified within Tn*7* and related transposons has not been characterised fully, but is known to comprise a pseudo-integrase gene (Pelletier and Roy, 1990; H-K. Young and S. Rosser, unpublished data; L. Sundström, personal communication) and up to three genetic cassettes in the central variable region (Sundström et al., 1991). Full characterisation of the 3' end of the element has not yet been achieved.

8.5.2 Trimethoprim Resistance in *Acinetobacter* spp.

Several Gram-negative bacteria, including *Acinetobacter* spp., are inherently resistant to low levels (MICs of 16–32 mg/L) of trimethoprim (Goldstein et al., 1983). High-level (MIC of >1000 mg/L) trimethoprim resistance in a clinical isolate of *Acinetobacter* (strain BM25000) was first reported by Goldstein et al. (1983). The resistance gene was located on a 167-kb self-transferable plasmid, pIP1031, which also encoded resistance to ampicillin, chloramphenicol, kanamycin, streptomycin and sulphamethoxazole. The biochemical mechanism and nature of the resistance gene were not determined, but hybridisation results indicated that trimethoprim resistance was not mediated by a type Ia DHFR or a transposon related in sequence to that of Tn*7*. Transferable high-level trimethoprim resistance was also reported by Chirnside et al. (1985) in a single strain of *Acinetobacter*. The 112-kb plasmid responsible carried resistance to sulphamethoxazole in addition to trimethoprim, but no attempt was made to characterise the *dhfr* gene involved.

High-level trimethoprim resistance is now common amongst multiresistant *Acinetobacter* strains responsible for nosocomial infections in France (Muller-Serieys et al., 1989) and Chile (G. Gonzalez, personal communication). Recent studies of the genetic basis of high-level trimethoprim resistance in Chilean isolates of *A. baumannii* showed that the strains could be classified into three groups according to their resistance phenotypes. All 30 isolates tested expressed resistance to ampicillin, cephaloridine, cefuroxime, ceftazidime, gentamicin and streptomycin in addition to trimethoprim resistance. Eleven isolates were additionally resistant to chloramphenicol and sulphamethoxazole, and 17 isolates were additionally resistant to amikacin, chloramphenicol and sulphamethoxazole. All 30 isolates were tested initially for the presence of transferable trimethoprim resistance by conjugation with rifampicin-resistant mutants of *E. coli, P. aeruginosa* and *Acinetobacter* strain BD413 in both liquid and solid phase matings. However, no transfer of trimethoprim resistance could be demonstrated at either 37 or 25°C. Attempts to mobilise trimethoprim resistance with the *inc*P group plasmid RP4 were also unsuccessful, despite the relatively high frequency of retransfer of RP4 from *Acinetobacter* to *E. coli* (2×10^{-3} transconjugants/donor cell). Hybridisation of total cellular DNA from 13 of the strains with a *dhfrIa* gene probe indicated the presence of this gene within the chromosome of each isolate. Subsequent positive hybridisation of chromosomal DNA preparations with probes specific for the integrase gene and the *tnsE* transposition gene suggested the presence of a Tn*7*-like transposon. In order to characterise the genetic surroundings of the *dhfrIa* gene further, and to extend the analysis of

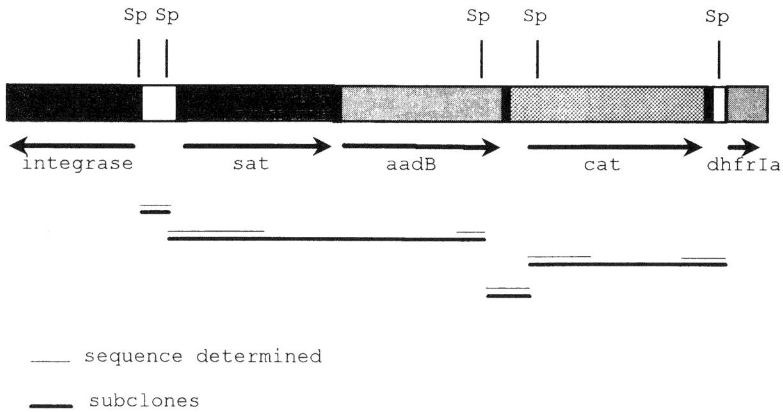

FIGURE 8.3
Structure of the 2.5-kb PCR product obtained from *Acinetobacter* strain ACE146.

trimethoprim resistance to all 30 highly resistant strains, each strain was subjected to PCR amplification. Highly specific oligonucleotide primers complementary to the *dhfrIa* N-terminal sequence and the N-terminal sequence of the Tn*7 int* gene were employed. Recovery of a PCR product following amplification was taken as proof that both the *dhfrIa* and Tn*7 int* genes were present in the isolate, and the presence of a 700-bp PCR product, containing two internal *Sph*I restriction sites, was interpreted as confirmation of the occurrence of the *dhfrIa* gene within a classic Tn7 integron structure. Of the 30 strains tested, a total of 27 gave the expected 700-bp Tn7-derived product. Interestingly, repeated PCR amplifications revealed the occasional presence of a second PCR product, about 2.5-kb in size, in a significant number of strains (11/23). On four occasions, the 2.5-kb fragment was the only product visualised following electrophoresis of the PCR product, although it was more usual to recover both the small and large PCR fragments. No correlation between isolate resistance profile and amplification of the second product could be made, and restriction analysis of the 2.5-kb PCR products amplified from three strains (ACE46, ACE145 and ACE146), representing the three different resistance patterns, revealed them to be identical. Subcloning and partial sequence analysis of restriction fragments of the 2.5-kb product from strain ACE146 revealed the presence of three genetic cassettes inserted between the integrase gene and the *dhfrIa* trimethoprim resistance gene. Figure 8.3 shows the extent of the sequence determination and the arrangement of the cassettes. The cassette located adjacent to the integrase gene showed complete homology, over the 350 bases sequenced, with the 5'-end of the streptothricin resistance gene (*sat*) of Tn*1826*, a Tn*7*-like transposon which lacks the *dhfrIa* gene cassette. An *aadB* gene (180-bp of the C-terminal determined), which encodes the ANT(2")-Ia aminoglycoside-modifying enzyme responsible for gentamicin resistance, was identified downstream of the *sat* gene cassette, and this was followed by a *cat* gene (250-bp of the 5'-end and 80-bp of the 3'-end determined) showing close homology with the novel chloramphenicol acetyltransferase gene (*cat*) present in Tn*2424* (Meyer et al., 1983; Parent and Roy, EMBL accession no. M80188).

These preliminary results are particularly interesting because they suggest the presence of a Tn*7*-related integron structure which is active in integration and deletion of resistance gene cassettes. Previous analysis of the integrase genes within Tn*7*-like transposons such as Tn*7* (Pelletier and Roy, 1990), Tn*4132* (H.K. Young and S. Rosser, unpublished results), Tn*1826* and Tn*1825* (L. Sundström, personal communication) has revealed the presence of an ochre stop codon interrupting translation of the integrase transcript, thus rendering it nonfunctional. An alternative explanation for these observations might be that a Tn*21* integrase is present in the strains and is functioning *in trans* to mediate insertion and deletion of cassettes into the Tn*7*-like integron structure. Although this is feasible, since the Tn*21* integrase has been shown to function *in trans,* it is unlikely to be the case in the Chilean *Acinetobacter* strains, since all isolates gave a negative hybridisation result with DNA probes specific for the Tn*21* integrase gene.

The identification of a highly active integration system in *Acinetobacter,* most probably located within a chromosome-located Tn*7*-like transposon, is of serious concern. The highly efficient, site-specific transposition mechanism of Tn*7* enables it to insert into the chromosome of many species, including *Acinetobacter,* and provides the host strain with a mechanism whereby resistance genes can become located permanently within the bacterial chromosome. The identification of an active chromosomally located integron within *Acinetobacter* strains has even greater significance, since enterobacterial plasmids, the majority of which are thought to be unstable in *Acinetobacter* spp., can still be acquired by *Acinetobacter* strains within the clinical setting, and can function as vehicles for the delivery of resistance gene cassettes into the integron structure on the genome of this new host species.

8.6 CHLORAMPHENICOL RESISTANCE

Chloramphenicol is often disregarded by clinical microbiologists as the drug is considered to be too toxic for routine use. However, its use in the past has ensured that a surfeit of chloramphenicol resistance genes exist, and these now provide the opportunity to examine the epidemiology of resistance genes without the continuous interference of new selective pressures.

Elisha and Steyn (1991a) examined the dissemination of chloramphenicol resistance genes as a convenient model for the scattering of resistance genes in general. The 650-bp gene encoding chloramphenicol acetyl transferase I (CAT1) encodes a protein that transfers an acetyl group to the chloramphenicol molecule and thus renders it incapable of challenging the 50S ribosomal subunit. Plasmid pA10*cat2* carries the CAT1 structural gene (Rosenthal et al., 1983), and Elisha and Steyn (1991b) used a 520-bp *Hin*dIII-*Nco*I fragment from pA10*cat2* as a gene probe. When radiolabelled with [^{32}P]dCTP, this probe gave no hybridisation with the chloramphenicol-sensitive *A. baumannii* strain YON. When DNA extracts of the chloramphenicol-resistant *A. baumannii* strain SAK were challenged with the radioactive probe, hybridisation was observed with both chromosomal and plasmid DNA. The plasmid involved was 120-kb in size and also conferred resistance to sulphonamides, streptomycin and mercury. These results suggested that the CAT1 gene might be transposon-encoded and had improved its survival potential by locating in both replicons. Elisha and Steyn (1991b) extracted chromosomal DNA and digested it with *Hin*dIII. A resultant 11-kb fragment contained a 5.0-kb *Sph*I fragment that was subcloned into plasmid pUC18 and, from this, a 1.8-kb *Pst*I fragment was subcloned. This final fragment contained a functional CAT gene. In contrast, when the 120-kb plasmid was digested with *Hin*dIII, a 5.0-kb fragment carried the CAT gene. In other words, the plasmid and chromosomal CAT determinants were located on different segments of DNA. When the DNA from the chromosome was sequenced, the 5' end of the CAT gene was linked to IS*1*, whereas the 3' end was linked to a 38-bp inverted repeat and part of the *tnpA* gene of Tn*21*.

A similar linkage has been reported for the CAT gene in the r-region of plasmid NR1. This plasmid belongs to the *inc*FII group, and the r-region is flanked by single copies of direct repeats of IS*1* so that it is transposable as Tn*2670*. It also contains a second transposon, Tn*21*, which encodes resistance genes other than chloramphenicol. As in the *Acinetobacter* DNA, the CAT gene on Tn*2670* in plasmid NR1 is located between IS*1a* and one of the inverted repeats of Tn*21* (Womble and Rownd, 1988). Therefore, Elisha and Steyn (1991b) concluded that the CAT gene in *Acinetobacter* strain SAK was also contained within a transposon, possibly similar to Tn*2670*, which had integrated into the host chromosome. Examination of the plasmid in the same strain suggested that it was not NR1, although the region around the CAT gene might be related. Elisha and Steyn (1991b) speculated that strain SAK might have acquired the putative transposon from a transient *inc*FII plasmid, since *inc*FII plasmids do not appear to be stable in *Acinetobacter* strains (Chopade et al., 1985), but may be present for sufficient time to allow transposition. Alternatively, the strain could have acquired a stable plasmid which had the r-region transposed on to it previously. A series of plasmids of different incompatibility groups have been transferred into *Acinetobacter* from *E. coli,* although the reverse transfer is often less frequent (Chopade et al., 1985).

Elisha and Steyn (1991b) argued that the resistances encoded by this transposon might be of little clinical significance. However, Tn*21* contains recombinational-hot-spots into which additional antibiotic resistance genes can insert. Wohlleben et al. (1989) described Tn*21* as a natural expression cassette for antibiotic resistance genes. Its presence in *Acinetobacter,* albeit flanking genes of little clinical importance, is cause for concern as resistance genes of considerable clinical significance could be acquired easily.

Another feature of resistance genes which are flanked by direct repeats of insertion sequences is that they can be amplified by a recombinational event between the insertion sequences. Womble and Rownd (1988) found that the r-region of plasmid NR1 could be amplified to form a poly-r-region which radically increased expression of the genes located within it and raised the MIC of those antibiotics. Thus, although chloramphenicol resistance is not important clinically in its own right, study of the CAT gene has revealed some of the possible mechanisms for the acquisition of foreign resistance genes by *Acinetobacter* spp.

8.7 OTHER RESISTANCE MECHANISMS

A major purpose of the bacterial membrane is to serve as a selective permeability barrier that controls the import and export of small molecular weight compounds. As described earlier, the water-filled proteinaceous diffusion channels, known as porins, are less numerous and smaller in *Acinetobacter* spp. than they are in other species. Such channels allow the cell to provide an effective barrier against hydrophobic compounds (Nikaido, 1979). There is evidence that *Acinetobacter* spp. possess a functional hydrophobic pathway (Borneleit and Kleber, 1991). The protection that these bacteria require in their normal environments of soil, water, skin and mucous membranes is quite different to enterobacteria. In addition, Baumann et al. (1968) demonstrated that many acinetobacters are capable of incorporating large hydrophobic metabolites, such as long-chain fatty acids, through the outer membrane. Borneleit et al. (1990; 1991) have shown that, in contrast to *E. coli* and *Salmonella typhimurium,* most *Acinetobacter* strains tested were sensitive to the hydrophobic antibiotic erythromycin. Members of the

Enterobacteriaceae had MICs of erythromycin in excess of 150 mg/L whereas, except for one strain, the *Acinetobacter* strains tested had MICs of erythromycin in the range of 1.2–50 mg/L, with a mean of 10 mg/L. Strains tested were also found to be relatively susceptible to novobiocin (MICs of 50–100 mg/L) and gentian violet (MICs of ≤1.5 mg/L).

8.8 CONCLUSIONS

Acinetobacter spp. must be considered as one of the greatest threats to the successful control of nosocomial infection. This threat does not come from any inherent virulence characteristics; indeed, these species are probably rather less virulent than bacteria associated previously with hospital infection. The danger resides in their capability to acquire and develop antibiotic-resistant genes, to the extent that these genes themselves could be considered as virulence factors. In particular, as surgical procedures advance and require prolonged immunosuppression of the patient, more reliance will be placed on antibiotics. Those genera that have the capability to develop resistance to the radical antibiotics used will prosper and predominate. Already, the successful control of *Acinetobacter* spp. by 4-quinolones, cephalosporins and perhaps even the carbapenems looks under threat. It is not clear how *Acinetobacter* spp. will be controlled once complete resistance to these three groups of antibacterial drugs has developed. Perhaps, though unlikely, a new group of antibiotics that will eradicate *Acinetobacter* spp. will be developed, although other 'environmental' bacteria may then take their place. The lessons that *Acinetobacter* spp. are teaching us should be assimilated by all clinicians, for if we are unable ultimately to control *Acinetobacter* spp., the overall struggle against resistance is in danger of being lost.

REFERENCES

Ambler, R.P. (1980). The structure of ß-lactamases. *Philos. Trans. R. Soc. London, Ser. B.* 289, 321–331.

Amyes, S.G.B. and Towner, K.J. (1990). Trimethoprim resistance; epidemiology and molecular aspects. *J. Med. Microbiol.* 31, 1–19.

Amyes, S.G.B., Towner, K.J. and Young, H.-K. (1992). Classification of plasmid-encoded dihydrofolate reductases conferring trimethoprim resistance. *J. Med. Microbiol.* 36, 1–3.

Barg, N. (1988). Construction of a probe for the aminoglycoside 3-V-acetyltransferase gene and detection of the gene among endemic clinical isolates. *Antimicrob. Agents Chemother.* 32, 1834–1838.

Barth, P.T., Datta, N., Hedges, R.W. and Grinter, N.J. (1976). Transposition of a deoxyribonucleic acid sequence encoding trimethoprim and streptomycin resistance from R483 to other replicons. *J. Bacteriol.* 125, 800–810.

Bauernfeind, A. (1986). Classification of ß-lactamases. *Rev. Infect. Dis.* 8 (Suppl. 5), S470–S481.

Baumann, P., Doudoroff, M. and Stanier, R.Y. (1968). A study of the *Moraxella* group II Oxidase-negative species (genus *Acinetobacter*). *J. Bacteriol.* 95, 1520–1541.

Berg, D., Davies, J., Allet, B. and Rochaix, J.D. (1975). Transposition of R factor genes to bacteriophage lambda. *Proc. Natl. Acad. Sci. U.S.A.* 72, 3628–3632.

Bergogne-Bérézin, E., Zechovsky, N., Piechaud, M., Vieu, J.F. and Bordini, A. (1971). Sensibilité aux antibiotiques de 240 souches de "*Moraxella*" oxydase-négative (*Acinetobacter*) isolées d'infections hopitalières. Evolution sur 3 ans. *Pathol. Biol.* 19, 981–990.

Bergogne-Bérézin, E., Joly, M.-L., Moreau, N. and Groffic, F. (1980). Aminoglycoside modifying enzymes in clinical isolates of *Acinetobacter calcoaceticus. Curr. Microbiol.* 4, 361–364.

Bissonnette, L. and Roy, P.H. (1992). Characterisation of InO of *Pseudomonas aeruginosa* plasmid pVS1, an ancestor of integrons of multiresistance plasmids and transposons of Gram-negative bacteria. *J. Bacteriol.* 174, 1248–1257.

Blechschmidt, B., Borneleit, P., Emanouilidis, I., Leihmann, W. and Kleber, H.-P. (1989). Extracellular location of a ß-lactamase produced by *Acinetobacter calcoaceticus. Appl. Microbiol. Biotechnol.* 32, 85–89.

Borneleit, P. and Kleber, H.-P. (1991). The outer membrane of *Acinetobacter:* structure function relationships. In *The Biology of* Acinetobacter, pp. 259–271. K.J. Towner, E. Bergogne-Bérézin and C.A. Fewson (Eds). Plenum Press, New York.

Borneleit, P., Binder, H. and Kleber, H.-P. (1990). Outer membrane vesiculation of *Acinetobacter calcoaceticus. Acta Biotech.* 10, 117–123.

Bouvet, P.J.M. and Grimont, P.A.D. (1987). Identification and biotyping of clinical isolates of *Acinetobacter. Ann. Microbiol. Inst. Pasteur* 28, 569–578.

Bush, K. (1989). Classification of ß-lactamases: Groups 1, 2a, 2b and 2b'. *Antimicrob. Agents Chemother.* 33, 264–270.

Chaslus-Dancla, E., Glupezynski, Y., Gerbaud, G., Lagorce, M., Lafont, J.P. and Courvalin, P. (1989). Detection of apramycin resistant Enterobacteriaceae in hospital isolates. *FEMS Microbiol. Lett.* 61, 261–266.

Chirnside, E.D., Emmerson, A.M. and Smith, J.T. (1985). A follow-up survey of transferable plasmid-encoded trimethoprim resistance in a general hospital (1975–1983). *J. Antimicrob. Chemother.* 16, 419–434.

Chopade, B.A., Wise, P.J. and Towner, K.J. (1985). Plasmid transfer and behaviour in *Acinetobacter calcoaceticus* EBF65/65. *J. Gen. Microbiol.* 131, 2805–2811.

Collis, C.M. and Hall, R.M. (1992). Site-specific deletion and rearrangement of integron insert genes catalysed by the integron DNA integrase. *J. Bacteriol.* 174, 1574–1585.

Collis, C.M., Grammaticopoulos, G., Briton, J., Stokes, H.W. and Hall, R.M. (1993). Site-specific insertion of gene cassettes into integrons. *Mol. Microbiol.* 9, 41–52.

Craig, N. (1989). Transposon Tn7. In *Mobile DNA*, pp. 211–225. D.E. Berg and M.M. Howe (Eds). American Society for Microbiology, Washington, D.C.

Devaud, M., Kayser, F.H. and Bachi, B. (1982). Transposon-mediated antibiotic resistance in *Acinetobacter* strains. *Antimicrob. Agents Chemother.* 22, 323–329.

Dowding, J.E. (1979). A novel aminoglycoside-modifying enzyme from a clinical isolate of *Acinetobacter. J. Gen. Microbiol.* 110, 239–241.

Du Bois, S.K., Marriott, M.S. and Amyes, S.G.B. (1995). TEM- and SHV-derived extended-spectrum ß-lactamases: relationship between selection, structure and function. *J. Antimicrob. Chemother.* 35, 7–22.

Elisha, B.G. and Steyn, L.M. (1989). Aminoglycoside-O-phosphotransferase (APH(3")) activity in a clinical isolate of *Acinetobacter calcoaceticus*. *S. Afr. Med. J.* 75, 220–222.

Elisha, B.G. and Steyn, L.M. (1991a). The use of molecular techniques for the location and characterisation of antibiotic resistance genes in clinical isolates of *Acinetobacter*. In *The Biology of* Acinetobacter, pp. 133–148. K.J. Towner, E. Bergogne-Bérézin and C.A. Fewson (Eds). Plenum Press, New York.

Elisha, B.G. and Steyn, L.M. (1991b). Identification of an *Acinetobacter baumannii* gene region with sequences and organizational similarity to Tn*2670*. *Plasmid* 25, 96–104.

Goldstein, F.W., Labingne-Roussel, A., Gerbaud, G., Carlier, C., Collatz, E. and Courvalin, P. (1983). Transferable plasmid mediated antibiotic resistance in *Acinetobacter*. *Plasmid* 10, 138–147.

Gomez-Lus, R., Larrad, L., Rubio-Calvo, M.C., Navarro, M. and Asierra, M.P. (1980). AAC(3) and AAC(6') enzymes produced by R-plasmids isolated in general hospital. In *Antibiotic Resistance*, pp. 295–303. S. Mitsuhashi, L. Rosival and V. Krcmery (Eds). Springer-Verlag, Berlin.

Hall, R.M., Brookes, D.E. and Stokes, H.W. (1991). Site-specific insertion of genes into integrons: role of the 59-base element and the determination of the recombination crossover point. *Mol. Microbiol.* 5, 1941–1959.

Heikkila, E., Renkonen, O.V., Sunila, R., Uurasmaa, P. and Huovinen, P. (1990). The emergence and mechanisms of trimethoprim resistance in *Escherichia coli* isolated from outpatients in Finland. *Antimicrob. Agents Chemother.* 25, 275–283.

Hikida, M., Yoshida, M., Mitsuhashi, S. and Inoue, M. (1989). Purification and properties of a cephalosporinase from *Acinetobacter calcoaceticus*. *J. Antibiot.* 42, 123–126.

Hood, J. (1991). The Chromosomal ß-Lactamases of the Genus Acinetobacter. PhD. Thesis, University of Edinburgh, Scotland.

Hood, J. and Amyes, S.G.B. (1989). A novel method for the identification and distinction of ß-lactamases of the genus *Acinetobacter*. *J. Appl. Bacteriol.* 67, 157–163.

Hood, J. and Amyes, S.G.B. (1991). The chromosomal ß-lactamases of the genus *Acinetobacter*: enzymes which challenge our imagination. In *The Biology of* Acinetobacter, pp. 117–132. K.J. Towner, E. Bergogne-Bérézin and C.A. Fewson (Eds). Plenum Press, New York.

Joly-Guillou, M.-L. and Bergogne-Bérézin, E. (1990). Présence d'une beta-lactamase à spretre élargi chez *Acinetobacter baumannii*. *Presse Med.* 19, 672–673.

Joly-Guillou, M.-L., Bergogne-Bérézin, E. and Moreau, N. (1987). Enzymatic resistance to ß-lactams and aminoglycosides in *Acinetobacter calcoaceticus*. *J. Antimicrob. Chemother.* 20, 773–775.

Joly-Guillou, M.-L., Bergogne-Bérézin, E. and Philippon, A. (1988). Distribution of ß-lactamases and phenotype analysis in clinical strains of *Acinetobacter calcoaceticus*. *J. Antimicrob. Chemother.* 22, 597–604.

Lambert, T., Gerbaud, G. and Courvalin, P. (1988). Transferable amikacin resistance in *Acinetobacter* spp. due to a new type of 3'-aminoglycoside phosphotransferase. *Antimicrob. Agents Chemother.* 32, 15–19.

Lambert, T., Gerbaud, G., Bouvet, P., Vieu, J.-F. and Courvalin, P. (1990). Dissemination of amikacin resistance gene *aphA6* in *Acinetobacter* spp. *Antimicrob. Agents Chemother.* 34, 1244–1248.

Lambert, T., Gerbaud, G., Galimand, M. and Courvalin, P. (1993). Characterisation of *Acinetobacter haemolyticus aac(6')-Ig* gene encoding an aminoglycoside *6'-N*-acetyltransferase which modifies amikacin. *Antimicrob. Agents Chemother.* 37, 2093–2100.

Licitra, C.M., Brooks, R.G., Terry, P.M., Shaw, K.J. and Hare, R.S. (1989). Use of plasmid analysis and determination of aminoglycoside-modifying enzymes to characterise isolates from an outbreak of methicillin-resistant *Staphylococcus aureus. J. Clin. Microbiol.* 27, 2535–2538.

Lovering, A.M., Bywater, M.J., Holt, H.A., Champion, H.M. and Reeves, D.S. (1988). Resistance of bacterial pathogens to four aminoglycosides and six other antibacterials and prevalence of aminoglycoside modifying enzymes in 20 U.K. centres. *J. Antimicrob. Chemother.* 22, 823–839.

Martin, P., Jullien, E. and Courvalin, P. (1988). Nucleotide sequence of *Acinetobacter baumannii* aphA-6 gene: evolutionary and functional implications of sequence homologies with nucleotide-binding proteins, kinases and other aminoglycoside-modifying enzymes. *Mol. Microbiol.* 2, 615–625.

Masuda, G., Tomioka, S. and Hasegawa, M. (1976). Detection of ß-lactamase production in Gram-negative bacteria. *J. Antibiot.* 29, 662–664.

Matthew, M. and Harris, A.M. (1976). Identification of ß-lactamase by analytical isoelectric focusing: correlation with bacterial taxonomy. *J. Gen. Microbiol.* 94, 55–67.

Matthew, M., Harris, A.M., Marshall, M.J. and Ross, G.W. (1975). The use of analytical isoelectric focusing for detection and identification of ß-lactamases. *J. Gen. Microbiol.* 88, 169–178.

Medeiros, A.A. (1984). ß-Lactamase. *Br. Med. Bull.* 40, 18–27.

Medeiros, A.A., Hare, R., Papa, E., Adam, C. and Miller, G.H. (1985). Gram-negative bacilli resistant to third-generation cephalosporins: ß-lactamase characterisation and susceptibility to Sch34343. *J. Antimicrob. Chemother.* 15 (Suppl. C), 119–132.

Meyer, J.F., Nies, B.A., Kratz, J. and Wiedemann, B. (1983). Amikacin resistance mediated by multiresistance transposon Tn*2424*. *J. Bacteriol.* 155, 755–760.

Minami, S., Yotsuji, A., Inoue, M. and Mitsuhashi, S. (1980). Induction of ß-lactamase by various ß-lactam antibiotics in *Enterobacter cloacae. Antimicrob. Agents Chemother.* 18, 382–285.

Morohoshi, T. and Saito, T. (1977). ß-Lactamase and ß-lactam antibiotics resistance in *Acinetobacter anitratum. J. Antibiot.* 30, 969–973.

Muller-Serieys, C., Lesquoy, J.B., Perez, E., Fichelle, A., Bourjeois, B., Joly-Guillou, M.-L. and Bergogne-Bérézin, E. (1989). Infections nocomiales à Acinetobacter: épidémiologie et difficultés thérapeutiques. *Presse Med.* 18, 107–110.

Murray, B.E. and Moellering, R.C. (1979). Aminoglycoside-modifying enzymes among clinical isolates of *Acinetobacter calcoaceticus* subsp. *anitratus (Herellea vaginicola).* Explanation for high-level aminoglycoside resistance. *Antimicrob. Agents Chemother.* 15, 190–199.

Murray, B.E. and Moellering, R.C. (1980). Evidence of plasmid-mediated production of aminoglycoside-modifying enzymes not previously described in *Acinetobacter. Antimicrob. Agents Chemother.* 17, 30–36.

National Committee for Clinical Laboratory Standards (1990). *Methods for Dilution Antimicrobial Susceptibility Tests for Bacteria that Grow Aerobically*, 2nd ed. National Committee for Clinical Laboratory Standards, Villanova, PA.

Neu, H.C. (1986). Antibiotic inactivating enzymes and bacterial resistance. In *Antibiotics in Laboratory Medicine*, pp. 757–789. V. Lorian (Ed.). Williams & Wilkins, Baltimore.

Neu, H.C. (1988). Bacterial resistance to fluoroquinolones. *Rev. Infect. Dis.* 10 (Suppl. 1), S57–S63.

Ng, E.Y.W., Trucksis, M. and Hooper, D.C. (1994). Quinolone resistance mediated by *norA*: Physiologic characterisation and relationship to *flqB*, a quinolone resistance locus on the *Staphylococcus aureus* chromosome. *Antimicrob. Agents Chemother.* 38, 1345–1355.

Nikaido, H. (1979). Permeability of the outer membrane of bacteria. *Angew. Chemie.* 18, 337–350.

O'Brien, T.F., del Pilar Pla, M., Mayer, K.H., Kishi, H., Gilleece, E., Syvanen, M. and Hopkins, J.D. (1985). Intercontinental spread of a new antibiotic resistance gene on an epidemic plasmid. *Science* 203, 87–88.

Obara, M. and Nakae, T. (1991). Mechanisms of resistance to ß-lactam antibiotics in *Acinetobacter calcoaceticus*. *J. Antimicrob. Chemother.* 28, 791–800.

Oka, A., Sugisak, H. and Takanami, M. (1981). Nucleotide sequence of the kanamycin resistance transposon Tn*903*. *J. Mol. Biol.* 14, 217–226.

Olsson, L. and Laas, T. (1981). Isoelectric focusing in agarose under denaturing conditions. *J. Chromatogr.* 215, 373–378.

Olsson, L. and Tooke, N.E. (1988). PAGE of basic proteins. In *Phastsystem Application File No. 300*. Phamacia, Uppsala.

Parr, T.R., Moore, L.V. and Hancock, R.E.W. (1987). Role of porins in intrinsic antibiotic resistance of *Pseudomonas cepacia*. *Antimicrob. Agents Chemother.* 31, 121–123.

Paton, R.H., Miles, R.S., Hood, J. and Amyes, S.G.B. (1993). ARI-1: Beta-lactamase-mediated imipenem resistance in *Acinetobacter baumannii*. *Int. J. Antimicrob. Agents.* 2, 81–88.

Paul, G., Joly-Guillou, M.-L., Bergogne-Bérézin, E., Nevot, P. and Philippon, A. (1989). Novel carbenicillin-hydrolysing ß-lactamase (CARB-5) from *Acinetobacter calcoaceticus* var *anitratus*. *FEMS Microbiol. Lett.* 59, 45–50.

Paulsen, I.T., Littlejohn, T.G., Radstrom, P., Sundström, L., Sköld, O., Swedberg, G. and Skurray, R.A. (1993). The 3' conserved segment of integrons contains a gene associated with multidrug resistance to antiseptics and disinfectants. *Antimicrob. Agents Chemother.* 37, 761–768.

Pelletier, A.H. and Roy, P.H. (1990). Bacterial transposon Tn*7* codes for an integrase pseudogene. In *Abstr. Annu. Meet. Am. Soc. Microbiol.* American Society for Microbiology, Washington, D.C.

Philippon, A.M., Paul, G.C. and Nevot, P.A. (1980). Synergy of clavulanic acid with penicillins against ampicillin- and carbenicillin-resistant Gram-negative organisms, related to their type of ß-lactamase. In *Current Chemotherapy and Immunotherapy* (Proc. 11th Int. Congr. Chemother. 19th Intersci. Conf. Antimicrob. Agents Chemother.) American Society for Microbiology, Washington, D.C.

Piddock, L.J.V., Panchal, S., Inglis, T.J.L. and Hawkey, P.M. (1992). High-level quinolone resistance in *Pseudomonas aeruginosa*. *J. Antimicrob. Chemother.* 30, 229–231.

Piddock, L.J.V. and Wise, R. (1986). Properties of the penicillin-binding proteins of four species of the genus *Bacteroides*. *Antimicrob. Agents Chemother.* 29, 825–832.

Quibell, K.J., Sato, K., Osada, Y. and Amyes, S.G.B. (1993). DNA gyrases in 4-quinolone resistant *Pseudomonas aeruginosa*. In *Abstr. 6th Eur. Congr. Clin. Microbiol. Infect. Dis. (Seville)*.

Robillard, N.J. (1990). Broad-host-range gyrase A gene probe. *Antimicrob. Agents Chemother.* 34, 1889–1894.

Rosenbusch, J.P. (1974). Characterization of the major envelope protein from *Escherichia coli*. *J. Biol. Chem.* 249, 8019–8029.

Rosenthal, N., Kress, M., Gruss, P. and Khoury, G. (1983). BK viral enhancer element and a human cellular homolog. *Science* 222, 749–755.

Saravis, C.A. and Zamcheck, N. (1979). Isoelectric focusing in agarose. *J. Immunol. Methods.* 29, 91–96.

Sato, K. and Nakae, T. (1991). Outer membrane permeability of *Acinetobacter calcoaceticus* and its implication in antibiotic resistance. *J. Antimicrob. Chemother.* 28, 33–45.

Scaife, W., Young, H.-K., Paton, R.H. and Amyes, S.G.B. (1995). Transferable imipenem resistance in *Acinetobacter* species from a clinical source. *J. Antimicrob. Chemother.* 36, 585–586.

Shannon, K.P., Phillips, I. and King, B.A. (1978). Aminoglycoside resistance among Enterobacteriaceae and *Acinetobacter* spp. *J. Antimicrob. Chemother.* 4, 131–142.

Shaw, K.J., Hare, R.S., Sabatelli, F.J., Rizzo, M., Cramer, C.A., Naples, L., Kocal, S., Munayyer, H., Mann, P., Miller, G.H., Verbist, L., Landuyt, H.V., Glupcaynski, Y., Catalano, M. and Woloj, M. (1991). Correlation between aminoglycoside resistance profiles and DNA hybridization of clinical isolates. *Antimicrob. Agents Chemother.* 35, 2253–2261.

Shaw, K.J., Rather, P.N., Hare, R.S. and Miller, G.H. (1993). Molecular genetics of aminoglycoside resistance genes and familial relationships of the aminoglycoside-modifying enzymes. *Microbiol. Rev.* 57, 138–163.

Sirot, J., Chanal, C., Petit, A., Sirot, D., Labia, R. and Gerbaud, G. (1988). *Klebsiella pneumoniae* and other Enterobacteriaceae producing novel plasmid-mediated beta-lactamases markedly active against third generation cephalosporins. *Rev. Infect. Dis.* 10, 850–859.

Soussy, C.J., Wolfson, J.S., Ng, E.Y. and Hooper, D.C. (1993). Limitations of the plasmid complementation test for determination of quinolone resistance (QR) due to changes in the gyrase A protein and identification of a conditional locus. In *Program Abstr. 33rd Intersci. Conf. Antimicrob. Agents Chemother. (New Orleans).*

Steen, R. and Sköld, O. (1985). Plasmid-borne or chromosomally mediated resistance by Tn*7* is the most common response to ubiquitous use of trimethoprim. *Antimicrob. Agents Chemother.* 27, 933–937.

Stokes, H.W. and Hall, R.M. (1989). A novel family of potentially mobile DNA elements encoding site-specific gene integration functions: integrons. *Mol. Microbiol.* 3, 1669–1683.

Sundström, L. and Sköld, O. (1990). The *dhfrIa* trimethoprim resistance gene of Tn7 can be found at specific sites in other genetic surroundings. *Antimicrob. Agents Chemother.* 34, 642–650.

Sundström, L., Roy, P.H. and Sköld, O. (1991). Site-specific insertion of three structural gene cassettes in transposon Tn*7*. *J. Bacteriol.* 173, 3025–3028.

Sykes, R.B. and Matthews, M. (1976). The ß-lactamases of Gram-negative bacteria and their role in resistance to ß-lactam antibiotics. *J. Antimicrob. Chemother.* 2, 115–157.

Sykes, R.B. and Smith, J.T. (1979). Biochemical aspects of ß-lactamases. In *ß-Lactamase*, pp. 369–410. J.M.T. Hamilton-Miller and J.T. Smith (Eds). Academic Press, London.

Tenover, F.C., Gootz, T.D., Gordon, K.P., Tompkins, L.S., Young, S.A. and Plorde, J.J. (1984). Development of a DNA probe for the structural gene of the 2"-O-adenyltransferase aminoglycoside-modifying enzyme. *J. Infect. Dis.* 150, 678–687.

Tenover, F.C., Phillips, K.L., Gilbert, T., Lockhart, P., O'Hare, P.J. and Plorde, J.J. (1989). Development of a DNA probe from the deoxyribonucleotide sequence of a 3-N-aminoglycoside acetyltransferase [ACC(3)-I] resistance gene. *Antimicrob. Agents Chemother.* 33, 551–559.

Towner, K.J. (1991). Plasmid and transposon behaviour in Acinetobacter. In *The Biology of* Acinetobacter, pp. 149–167. K.J. Towner, E. Bergogne-Bérézin and C.A. Fewson (Eds). Plenum Press, New York.

Towner, K.J., Young, H.-K., Thomson, C.J. and Amyes, S.G.B. (1990). Detection in the U.K. of trimethoprim-resistant *Escherichia coli* encoding the type V dihydrofolate reductase. *Eur. J. Clin. Microbiol. Infect. Dis.* 9, 149–150.

van der Klundert, J.A.M., Vliegenthart, J.S., van Doorn, E., Bongaerts, G.P.A., Molendijk, L. and Mouton, R.P. (1984). A simple method for the identification of aminoglycoside-modifying enzymes. *J. Antimicrob. Chemother.* 14, 339–348.

Vecoli, C., Prevost, F.E., Ververis, J.J., Medeiros, A.A. and O'Leary, G.P. (1983). Comparison of polyacrylamide and agarose gel thin layer isoelectric focusing for the characterisation of ß-lactamases. *Antimicrob. Agents Chemother.* **24**, 186–189.

Vila, J., Marcos, A., Marco, F., Abdalla, S., Vergara, Y., Reig, R., Gomez-Lus, R. and Deanta, T.J. (1993). In vitro antimicrobial production of beta-lactamases, aminoglycoside-modifying enzymes, and chloramphenicol acetyltransferase by and susceptibility of clinical isolates of *Acinetobacter baumannii. Antimicrob. Agents Chemother.* 37, 138–141.

Vila, J., Ruiz, J., Goni, P., Marcos, M.A. and Jiminez de Anta, M.T. (1994). Detection of *gyrA* gene mutations in fluoroquinolone resistant clinical isolates of *Acinetobacter baumannii*. In *Abstr. 3rd Int. Symp. Biol.* Acinetobacter *(Edinburgh)*.

Vliegenthart, J.S., Ketalaar-van Gaalen, P.A.G. and van de Klundert, J.A.M. (1989). Nucleotide sequence of the *aacC2* gene, a gentamicin resistance determinant involved in a hospital epidemic of multiply resistant members of the family Enterobacteriaceae. *Antimicrob. Agents Chemother.* 33, 1153–1159.

Wohlleben, W., Arnold, W., Bissonnette, L., Pelletier, A., Tanguay, A., Roy, P.H., Gamboa, G.C., Barry, G.F., Aubert, E., Davies, J. and Kagan, S.A. (1989). On the evolution of Tn*21*-like multiresistance transposons: sequence analysis of the gene *(aac(3)-I)* for gentamicin acetyltransferase-3-I (AAC(3)-I), another member of the Tn*21*-based expression cassette. *Mol. Gen. Genet.* 217, 202–208.

Womble, D.D. and Rownd, R.H. (1988). Genetic and physical map of plasmid NR1: comparison with other IncFII antibiotic resistance plasmids. *Microbiol. Rev.* 52, 433–451.

Yoshimura, F. and Nikaido, H. (1982). Permeability of *Pseudomonas aeruginosa* outer membrane to hydophilic solutes. *J. Bacteriol.* 152, 636–642.

Young, H.-K. and Hillyear, J.K. (1994). Trimethoprim resistance in urinary pathogens in northern Scotland: epidemic spread of a resistance plasmid encoding the type Ib trimethoprim-resistant dihydrofolate reductase. *J. Med. Microbiol.* 41, 343–348.

Young, S.A., Tenover, F.C., Gootz, T.D., Gordon, K.P. and Plorde, J.J. (1985). Development of two DNA probes for differentiating the structural genes of subclasses I and II of the aminoglycoside-modifying enzyme 3'-aminoglycoside phosphotransferase. *Antimicrob. Agents Chemother.* 27, 739–745.

Chapter 9

Nosocomial Infections Caused by *Acinetobacter* spp. — Therapeutic Problems

M. Wolff

CONTENTS

0-8493-9223-3/95/$0.00+$.50

9.1 INTRODUCTION

Although they can be responsible for occasional community-acquired infections (see Chapter 5), members of the genus *Acinetobacter* have emerged primarily as significant nosocomial pathogens, especially in intensive care units (ICUs). In the EPIIC study conducted in 17 European countries, *Acinetobacter* was isolated in 9.3% of ICU-acquired infections, particularly those involving blood and the lower respiratory tract (Wolff and Nicolas, 1994). A separate study performed in France showed similar results (Joly-Guillou et al., 1992). Because of its increasing resistance to many available antibiotics, *Acinetobacter* is nowadays one of the most difficult-to-treat pathogens. A retrospective study conducted in 89 ICUs in France showed that, although *Acinetobacter* "was a problem" in only one third of the ICUs surveyed, major therapeutic difficulties were encountered by >80% of physicians (Joly-Guillou et al., 1993a). This chapter reviews therapeutic approaches for *Acinetobacter* infections, with particular respect to treatment of severe nosocomial infections.

9.2 GENERAL CONSIDERATIONS

9.2.1 Clinical Impact of Appropriate Antibiotic Management

Early appropriate antibiotic treatment is able to reduce the fatality rate of Gram-negative bacteraemia by approximately 50%. In addition, the frequency with which shock develops is reduced, and even after the development of shock, patients with adequate antibiotic therapy have a better prognosis (Kreger et al., 1980). Since *Acinetobacter* is considered to be an organism of relatively low virulence, the role of early appropriate therapy has been a matter for debate. Tilley and Roberts (1994) studied the outcome of bacteraemia caused by *Acinetobacter* in 52 patients. The prognosis appeared to correlate more closely with the type of underlying disease and its severity than with other factors, including the appropriateness of the initial therapy. In another study of 27 patients with nosocomial *Acinetobacter* bacteraemia, 6 had polymicrobial sepsis that carried a higher mortality than 'pure' *Acinetobacter* bacteraemia (50% vs. 0%) (Smego, 1985). In contrast, in a case-control study, Fagon et al. (1993) found that ventilator-associated pneumonia caused by *Pseudomonas aeruginosa* or *Acinetobacter* spp. was responsible for an attributable mortality of 42.8%, compared to 27% in the overall population of patients with ventilator-associated pneumonia. In another control study performed during an outbreak of *A. baumannii* infections, the mortality attributable to respiratory tract infections was 50% (Kaul et al., 1994).

All of these studies indicate that empiric antimicrobial therapy active against *Acinetobacter* is justified when this pathogen is suspected on the basis of epidemiological surveillance and direct examination of clinical samples. However, treatment failure can result from choice of the wrong antimicrobial agent when the infecting *Acinetobacter* strain is multiresistant. The risk of inappropriate therapy is greater at the beginning of an outbreak when the susceptibility pattern is still unknown (Vandenbroucke-Grauls, 1993).

9.2.2 Rational Basis for Treatment of *Acinetobacter* Infections

The current antimicrobial susceptibility patterns of *Acinetobacter* spp. are discussed in detail in Chapter 7 and will only be considered briefly here.

In Vitro Susceptibility Data

Approximately 75% of *Acinetobacter* strains are now resistant to third-generation cephalosporins, including ceftazidime. This resistance is caused by the production of cephalosporinases (see Chapter 8). In contrast, ticarcillin is still active against 55% of strains in France. At present, resistance to imipenem remains rare among *Acinetobacter* strains, with <1% of strains showing significant resistance to this antibiotic. Amongst the aminoglycosides, netilmicin and amikacin have the greatest activity, with 60 and 45% of strains, respectively, remaining susceptible. Unfortunately, the vast majority of strains isolated from infections in France are now resistant to fluoroquinolones, including ciprofloxacin and ofloxacin (Joly-Guillou et al., 1993a).

The bactericidal effect of different regimens against strains of various resistance phenotypes was studied by Joly-Guillou et al. (1993b) with the killing curve technique. Imipenem was the most active compound against all strains, but synergy was demonstrated between ß-lactams and amikacin, compared with either antibiotic alone, although this synergy was lost if strains had a low level of resistance to ß-lactams (except ticarcillin) or aminoglycosides. A combination of sulbactam, which is intrinsically active against *Acinetobacter*, with ticarcillin and aminoglycosides showed the best bactericidal activity against strains which were multiresistant to ß-lactams.

9.2.3 General Recommendations

The therapeutic regimen for severe nosocomial Gram-negative infections usually consists of a combination of a ß-lactam with an aminoglycoside. There are several theoretical reasons which support the use of such a combination in this setting:

1. The advantage of a broad-spectrum regimen is provided for initial empiric therapy;
2. Rapid bactericidal therapy may be associated with a better outcome;
3. Combination therapy may limit the development of resistance during the course of treatment.

However, it is by no means clear that all of these factors are applicable for *Acinetobacter* infections. It has been demonstrated that patients with *P. aeruginosa* infections treated with combination therapy have a better outcome compared to patients receiving monotherapy (Hilf et al., 1989) but, to our knowledge, such data have not yet been published for *Acinetobacter* infections. Moreover, because of the overall increase in resistance levels, combination therapy is not always possible, especially when strains showing multiple resistance to ß-lactams are involved. Finally, the development of resistance to imipenem in strains of *Acinetobacter* during the actual course of therapy appears to be infrequent at the present time.

The initial empiric therapy for suspected cases of *Acinetobacter* infection should be guided by a knowledge of the local strain ecology and susceptibility patterns (both in the hospital as a whole and in the particular unit where the patient is based), and by the results of direct examination of appropriate samples. In units where the percentage of multiresistant strains is low, a combination of ceftazidime or ticarcillin with an aminoglycoside is usually effective. If the patient is hospitalised in a unit in which the prevalence of multiresistant strains is high, then imipenem associated with amikacin can be used. Following definitive identification of the strain causing the infection, therapy can be modified according to the results of susceptibility tests.

9.3 MANAGEMENT OF SPECIFIC INFECTIONS

9.3.1 Nosocomial Pneumonia

Parenteral Antibiotics

Treatment is often initiated before culture results are available. The empiric regimen is selected on the basis of examination of respiratory tract samples by Gram's stain. A diagnosis of *Acinetobacter* is suspected when Gram-negative diplococci or coccobacilli are seen. Since up to 20% of nosocomial pneumonias are polymicrobial, a broad-spectrum regimen should be initiated, usually the ß-lactam plus aminoglycoside combination which has become a standard empirical regimen for treating pneumonia caused by Gram-negative bacilli.

Knowledge of the usual susceptibility patterns of *Acinetobacter* strains acquired in each particular unit is crucial for the initial therapeutic decision. Definitive antimicrobial treatment can then be modified (or not) on the basis of subsequent antimicrobial susceptibility tests. In units where sensitive strains are normally isolated, a combination of ticarcillin (either alone or with clavulanate) and an aminoglycoside may be used, as suggested by several trials. For example, eight patients with nosocomial *Acinetobacter* pneumonia have been treated recently in Bichat and Argenteuil hospitals with ticarcillin (n = 5) or ticarcillin plus clavulanate (n = 3). Four patients had polymicrobial pneumonia involving seven pathogens (*P. aeruginosa*, three; methicillin-resistant *Staphylococcus aureus*, three; *Streptococcus* spp., one). Bacteriological diagnosis was arrived at by means of a protected-specimen brush technique. Three patients also received vancomycin, while amikacin was used in two cases. Bacteriological evaluation was possible in six cases: eradication was obtained in three patients, including two who were given ticarcillin as monotherapy; a ticarcillin-resistant strain emerged during the course of therapy in a further two cases, requiring the use of imipenem (Joly-Guillou et al., 1993c). In a multicentre open study, ticarcillin plus clavulanate, alone or in combination with an aminoglycoside, was administered to 61 patients with nosocomial infections, the majority of which were pneumonias. *Acinetobacter* was isolated in 19 cases, and the overall rate of success was 67% (Bismuth et al., 1989).

Unfortunately, the percentage of strains resistant to most ß-lactams is currently increasing, and imipenem is one of the last agents to retain activity against the vast majority of strains. Because of its extended spectrum, imipenem has become one of the first-line antimicrobial agents for treating pneumonias caused by Gram-negative bacilli. Peak concentrations of imipenem in human lung parenchyma reach 7 µg/g after a 1-g i.v. infusion (Loos et al., 1991). The efficacy of imipenem has been shown in a large number of clinical studies dealing with nosocomial respiratory tract infection (e.g., Cometa et al., 1994), including those caused by *Acinetobacter* spp. Among 16 patients receiving imipenem as monotherapy (1–4 g/day), *Acinetobacter* eradication was obtained in 14 cases (Acar, 1985; Dirksen et al., 1986; Giamarellou et al., 1986).

A recent multicentre, randomised, double-blind trial compared intravenous imipenem (1 g every 8 h) with ciprofloxacin (400 mg every

8 h) for the treatment of severe pneumonia in hospitalised patients. Only 11 (3%) of the 359 pneumonias studied were caused by *Acinetobacter*, with eradication in 5/6 and 4/5 cases in the ciprofloxacin and imipenem groups, respectively (Fink et al., 1994). In our unit, 82 ventilator-associated pneumonias were observed between 1990 and 1994, of which 15 (18%) were caused by *Acinetobacter* spp. Four of these 15 were polymicrobial. Imipenem was used in all but one patient, either alone (eight cases) or in combination with other antibiotics (six cases). Among patients who were given a combination, four received parenteral aminoglycoside, one received nebulised tobramycin, and one received polymyxin B. Eight patients died. Eradication was obtained in only four of the ten patients who could be evaluated microbiologically. No resistance to imipenem emerged during the course of therapy. One patient was treated with a combination of ticarcillin/clavulanate plus tobramycin; the clinical outcome was favourable despite the persistence of a susceptible strain in the trachea.

A major potential problem is that increased use of imipenem has recently led to the development of imipenem resistance in a few strains of *Acinetobacter* spp. An outbreak caused by an *Acinetobacter* strain resistant to imipenem has been reported in a surgical ICU in New York (Urban et al., 1993). Pneumonia or tracheobronchitis was diagnosed in nine patients. All were given a combination of ampicillin and sulbactam at a dosage of 3 g every 6 or 8 h. Bacterial eradication was obtained in five cases.

In contrast to the ß-lactams, the bactericidal effect of aminoglycosides is dependent on the peak concentration. Moore et al. (1984) found a significant relationship between peak plasma concentrations and the clinical outcome in 37 patients with Gram-negative pneumonia. In addition, high peak concentrations may limit the emergence of resistant variants.

Localised Antibiotic Therapy

Localised administration of aminoglycosides or polymyxin B by aerosolisation or direct instillation appears to be an attractive option for the treatment of pneumonia caused by Gram-negative bacilli, especially when multiresistant *Acinetobacter* is involved. This mode of administration can result in elevated and persistent concentrations of these antibiotics in bronchial secretions and in lung parenchyma, while plasma levels remain low. Le Conte et al. (1993) studied the initial lung distribution and pharmacokinetics of nebulised tobramycin in five mechanically ventilated patients without lung disease. Scintigraphy with radiolabelled antibiotic showed a homogenous distribution of the radioactivity in the whole lung, despite a significant deposition on the trachea and bronchi. In plasma, the peak concentrations were <0.1 mg/L. In another group of patients undergoing thoracic surgery, mean lung tissue concentrations 4 and 12 h after a 300-mg nebulisation of tobramycin were 5.5 and 3.61 μg/g, respectively (Le Conte et al., 1993). Baran et al. (1990) found very high and long-lasting concentrations of tobramycin in the epithelial lining fluid obtained by bronchoalveolar lavage after aerosolisation or direct instillation. A major concern with this route could be the potential heterogeneity in the pattern of deposition of particles within infected lungs. Previous studies have shown that deposition may be low in areas of consolidation or atelectasis. Guillaume et al. (1992) studied the pharmacokinetics of nebulised tobramycin in six mechanically ventilated patients with bacterial pneumonia, and found that the amount of tobramycin reaching the lung after aerosolisation, estimated by the bioavailable systemic dose, was 11% of the inhaled dose.

Aminoglycosides have also been administered by direct endotracheal instillation. This technique allows very high bronchial concentrations of the drug to be achieved. After the administration of 40 mg of gentamicin by endotracheal instillation, the bronchial levels of this drug are in the range of 700 μg/ml (Baran et al., 1975). Clinical trials have been performed in the setting of prophylactic rather than therapeutic localised administration. Endotracheal administration of aminoglycosides or polymyxin B to prevent pneumonia in mechanically ventilated patients was followed by the emergence of pathogens resistant to these antibiotics (Feeley et al., 1975; Levine et al., 1978). There is very limited experience of the therapeutic effectiveness of this route of administration in patients with established ventilator-associated pneumonia. Since many strains of *Acinetobacter* are now susceptible only to imipenem and polymyxin B, the efficacy and the safety of a combination of i.v. imipenem and nebulised polymyxin should be evaluated. Although some alteration in ventilatory functions has been observed in patients treated with aerosolised polymyxin, the endotracheal administration of aminoglycosides, either by direct instillation or aerosolisation, is usually well tolerated.

Prevention of Infection by Selective Decontamination of the Digestive Tract (SDD)

Timsit et al. (1993) have suggested that the digestive tract may be an important reservoir for *A. baumannii* in ICU patients. This finding could be a rationale for the use of SDD in order to prevent *Acinetobacter* respiratory tract infections. Polymyxin B, which is normally active against *Acinetobacter* spp., is usually included in most SDD regimens. A large number of studies have evaluated the efficacy of SDD and reported a significant decrease in colonisation and infection of the respiratory tract with Gram-negative bacilli, including *Acinetobacter* spp.

These optimistic results have been challenged by two recent studies. Lance-Sauders et al. (1994) reported results obtained with surveillance cultures during the course of a 2-year trial of the efficacy of SDD in an ICU. *Acinetobacter* spp. were the most frequently isolated Gram-negative bacilli causing respiratory tract infections in both placebo and SDD groups, and contributed to the overall limited efficacy of SDD. Similarly, a prospective study performed in two ICUs at Bichat Hospital in Paris examined the mode of *Acinetobacter* acquisition in patients receiving SDD with polymyxin B and erythromycin base. Chromosomal macrorestriction pulsed-field gel electrophoresis showed that patients developed more infections with ICU-acquired isolates than with their own strain. Therefore, SDD may not be an effective strategy for preventing ICU-acquired respiratory tract infections caused by *Acinetobacter* spp. since such infections appear to be mainly exogenous in origin (Joly-Guillou et al., 1994).

9.3.2 Bacteraemia

The respiratory tract and wounds are the main sources of nosocomial bacteraemia caused by *Acinetobacter* spp. Bacteraemia without an identifiable focus of infection is infrequent. In contrast to some other organisms, intravenous catheters are rarely involved (Tilley and Roberts, 1994). An active ß-lactam, associated if possible with an aminoglycoside, is the accepted regimen for the treatment of *Acinetobacter* bacteraemia. In addition, since the skin is the portal of entry of *Acinetobacter* in burn patients, skin disinfection is recommended in this setting.

9.3.3 Central Nervous System Infections

Acinetobacter is a not infrequent cause of meningitis or ventriculitis. Nearly all cases reported in adults have occurred after neurosurgical procedures or head trauma (Siegman-Igra et al., 1993). Treatment of such infections should be based upon the results of susceptibility tests and the level of penetration of active antibiotics into the cerebrospinal fluid (CSF). Since it can be difficult to achieve CSF levels that are tenfold higher than the MIC (as recommended by Decazes et al., 1983), the use of an intrathecal or intraventricular aminoglycoside may be necessary.

In our hospital, 8 of 22 patients included in an open study evaluating the efficacy of imipenem had ventriculitis or meningitis caused by *Acinetobacter* (Rouveix et al., 1986). *Acinetobacter* was associated with other pathogens (*Streptococcus mitis* and *Klebsiella pneumoniae*) in two cases, and was the sole pathogen in six cases. Imipenem was administered at a dosage of 2–4 g/day for 5–20 days. Intravenous and intrathecal or intraventricular aminoglycosides were used in three cases (tobramycin in two cases and amikacin in one case). Three patients were only given aminoglycosides by the intraventricular route. Imipenem was used as monotherapy in two cases. CSF sterilisation was achieved in all cases within 2–6 days (mean of 4 days). However, one patient relapsed after a 20-day course of imipenem associated with intravenous and local tobramycin, but was cured after a second 20-day course of the same regimen. The presence of a foreign body was probably responsible for the relapse. One of the eight patients died.

Serum and CSF imipenem levels were measured four times for two patients. In one patient, who was given 500 mg every 6 h, the serum peak level was 50 mg/L, and the CSF concentration was 2.6 mg/L at 150 min following the i.v. infusion. In the other patient, who was given 1000 mg every 6 h, serum peak levels ranged from 31–62 mg/L, and CSF levels from 1.4–3.8 mg/L at 150 min after the i.v. infusion. These CSF concentrations were two- to fourfold higher than the MIC for *Acinetobacter*. Both patients were given aminoglycosides by the i.v. and local routes.

Siegman-Igra et al. (1993) reviewed 25 cases of nosocomial *Acinetobacter* meningitis secondary to invasive procedures. For seven patients, the empirical therapy (ureidopenicillin plus an aminoglycoside) matched *in vitro* susceptibilities. For 16 patients, it was necessary to alter the regimen following *in vitro* susceptibility tests, in some cases to include the use of local amikacin injections. Unfortunately, details of the different regimens used were not reported by the authors. Two patients received inappropriate antibiotics and two patients died before receiving therapy. Eighteen of the 21 patients who were treated appropriately recovered from infection. Schönwald et al. (1989) used ciprofloxacin (200 mg every 12 h) to treat 20 patients with meningitis caused by Gram-negative bacilli, including three cases caused by *Acinetobacter*. All patients, including the latter, were cured. *Acinetobacter* could not be recovered from the CSF after therapy for 24 h, but CSF levels of ciprofloxacin were not reported. Segev et al. (1990) used i.v. pefloxacin at a dosage of 800 mg twice a day to treat four patients with *Acinetobacter* meningitis. Eradication was obtained in all four patients, with a mean CSF pefloxacin concentration of 8.8 mg/L.

Finally, some overall recommendations can be made for the treatment of *Acinetobacter* meningitis. Imipenem should be used for eradication of infections caused by multiresistant strains. Ticarcillin, ureidopenicillins or fluoroquinolones may be appropriate for use against susceptible strains. The combined use of an aminoglycoside administered by the i.v. and/or local route may not always be necessary, provided that high doses of ß-lactams or fluoroquinolones are used. Meropenem is a novel carbapenem antibiotic that is currently active against *Acinetobacter*, with CSF concentrations that range from 0.3–6.5 mg/L at 2.5–3.0 h following a 40 mg/kg infusion (Dagan et al., 1994). This agent may also be useful for the treatment of *Acinetobacter* meningitis caused by otherwise resistant strains.

9.4 CONCLUSIONS

Multiresistance may pose immense problems for the physician who has to treat severe nosocomial *Acinetobacter* infections. Initial reports of the emergence of imipenem-resistant strains are extremely worrying because of the limited therapeutic possibilities for eradicating these bacteria. In order to prevent the further emergence and spread of imipenem-resistant strains, ticarcillin with or without a ß-lactamase inhibitor should be used whenever possible. The development of new therapeutic molecules and their evaluation in animal models is an urgent requirement. In common with other multiresistant nosocomial pathogens, only the application of strict hygiene rules and other control of cross-infection measures will prevent the dissemination of *Acinetobacter* spp. within the hospital environment.

REFERENCES

Acar, J.F. (1985). Therapy for lower respiratory tract infections with imipenem-cilastatin: a review of worldwide experience. *Rev. Infect. Dis.* 7 (Suppl. 3), 513–517

Baran, D., Dachy, A. and Klastersky, J. (1975). Concentrations of gentamicin in bronchial secretions of children with cystic fibrosis or tracheotomy. *Int. Clin. Pharmacol.* 12, 336–341.

Baran, D., de Vuyst, P. and Ooms, H.A. (1990). Concentrations of tobramycin given by aerosol in the fluid obtained by bronchoalveolar lavage. *Respir. Med.* 83, 203–204.

Bismuth, F., Motin, J. et le Groupe Multicentrique Francais (1989). Utilisation du Claventin en antibiothérapie de première intention au cours d'infections nosocomiales survenant en réanimation. *Méd. Mal. Infect.* 19, 48–53.

Cometa, A., Baumgartner, J.D., Lew, D., Zimmerli, W., Pittet, D., Chopart, P., Schaad, U., Herter, C., Eggimann, P., Huber, O., Ricou, B., Suter, P., Auckenthaler, R., Chiolero, R., Bille, J., Scheidigger, C., Frei, R. and Glauser, M.P. (1994). Prospective randomized comparison of imipenem monotherapy with imipenem plus netilmicin for treatment of severe infections in nonneutropenic patients. *Antimicrob. Agents Chemother.* 38, 1309–1313.

Dagan, R., Velghe, L., Rodda, J.L. and Klugman, K.P. (1994). Penetration of meropenem into the cerebrospinal fluid of patients with inflamed meninges. *J. Antimicrob. Chemother.* 34, 175–179.

Decazes, J.M., Ernst, J.D. and Sande, M.A. (1983). Correlation of in vitro kill curves and kinetics of bacterial killing in CSF during ceftriaxone therapy of experimental *E. coli* meningitis. *Antimicrob. Agents Chemother.* 24, 963–967.

Dirksen, M.S.C., Wintermans, R.G.F., Boerema, J.B.J. and Jimbrere, J.S.F. (1986). Imipenem as monotherapy in the treatment of intensive care patients with severe infections. *J. Antimicrob. Chemother.* 18 (Suppl. E), 145–151.

Fagon, J.Y., Chastre, J., Hance, A., Montravers, Ph., Novara, A. and Gibert, C. (1993). Nosocomial pneumonia in ventilated patients: a cohort study evaluating attributable mortality and hospital stay. *Am. J. Med.* 94, 281–288.

Feeley, T.W., Dumoulin, G.C., Hedley-Whyte, J., Bushnell, L.S., Gilbert, J.P. and Feingold, D.S. (1975). Aerosol polymyxin and pneumonia in seriously ill patients. *N. Engl. J. Med.* 293, 471–475.

Fink, M.P., Snydman, D.R., Niederman, M.S., Leeper, K.V., Johnson, R.H., Heard, S.O., Wunderink, R.G., Caldwell, J.W., Schentag, J.J., Siami, G.A., Zameck, R.L., Haverstock, D.C., Reinhart, H.H., Echols, R.M. and the Severe Pneumonia Study Group (1994). Treatment of severe pneumonia in hospitalized patients: results of a multicenter, randomized, double-blind trial comparing intravenous ciprofloxacin with imipenem-cilastatin. *Antimicrob. Agents Chemother.* 38, 547–557.

Giamarellou, H., Sfikakis, P., Voutsinas, D., Galanakis, N. and Daikos, G. (1986). Evaluation of imipenem-cilastatin against nosocomial infections and multiresistant pathogens. *J. Antimicrob. Chemother.* 18 (Suppl. E), 175–179.

Guillaume, C., Faurisson, F. and Peytavin, G. (1992). Pharmacokinetics of nebulized tobramycin in mechanically ventilated patients with bacterial pneumonia. *Program Abstr. 32nd Intersci. Conf. Antimicrob. Agents Chemother. (Anaheim).*

Hilf, M., Yu, V.L., Sharp, J., Zuravleff, J.J., Korvick, J.A. and Muder, R.R. (1989). Antibiotic therapy for *Pseudomonas aerginosa* bacteraemia: outcome correlations in a prospective study of 200 patients. *Am. J. Med.* 87, 540–545.

Joly-Guillou, M.-L., Decré, D. and Bergogne-Bérézin, E. (1992). Infections nosocomiales à *Acinetobacter*. Surveillance épidémiologique hospitalière. *Bull. Epidémiol. Hebdominaine* vl, 211–212.

Joly-Guillou, M.-L, Decré, D., Wolff, M. and Bergogne-Bérézin, E. (1993a). Infections nosocomiales à *Acinetobacter*. Bilan d'une enquête rétrospective dans les unités de soins intensifs des CHU francais en 1991. *Bull. Epidémiol. Hebdominaine* x, 44–45.

Joly-Guillou, M.-L., Decré, D., Hermann, J.L. and Bergogne-Bérézin, E. (1993b). Bactericidal activity of beta-lactam inhibitors alone or associated to beta-lactams against strains of *Acinetobacter baumannii* according to resistance phenotypes; effects of combination with aminoglycosides. *Program Abstr. 33rd Intersci. Conf. Antimicrob. Agents Chemother. (New Orleans).*

Joly-Guillou, M.-L., Sollet, J.P., Varache, C. and Bergogne-Bérézin, E. (1992c). Infections dues à *Acinetobacter baumannii. Méd. Mal. Infect.* 23 (Spécial), 67–72.

Joly-Guillou, M.-L., Wolff, M., Decré, D., Calvat, S. and Bergogne-Bérézin, E. (1994). Colonization and infection with *Acinetobacter baumannii* in ICU patients receiving selective digestive decontamination: results of a case-control and molecular epidemiologic investigations. *Program Abstr. 34th Intersci. Conf. Antimicrob. Agents Chemother. (Orlando).*

Kaul, R., Cork, L., Dedier, H., Burt, J., Garcia, M., Krajden, M., Conly, J. (1994). A multi-year, multi-intensive care unit outbreak of *Acinetobacter baumannii*: molecular epidemiology and case-control study. *Program Abstr. 34th Intersci. Conf. Antimicrob. Agents Chemother. (Orlando).*

Kreger, B.E., Craven, D.E., Carling, O.C. and McCabe, W.R. (1980). Gram negative bacteraemia. IV. Re-evaluation of clinical features and treatment in 612 patients. *Am. J. Med.* 68, 344–355.

Lance-Sauders, G., Hammond, J.M.J., Potgieter, P.D., Plum, H.A. and Forder, A.A. (1994). Microbiological surveillance during selective decontamination of the digestive tract (SDD). *J. Antimicrob. Chemother.* 34, 529–544.

Le Conte, Ph., Potel, G., Peltier, P., Horeau, D., Caillon, J., Juvin, M.E., Kergueris, M.F., Bugnon, D. and Baron, D. (1993). Lung distribution and pharmacokinetics of aerosolized tobramycin. *Am. Rev. Respir. Dis.* 147, 1279–1282.

Levine, B.A., Petroff, P.A., Slade, C.L. and Pruitt, B.A. (1978). Prospective trials of dexamethasone and aerosolized gentamicin in the treatment of inhalation injury in the burned patients. *J. Trauma* 18, 188–193.

Loos, U., Friess, D. and Springmann, K. (1991). Pulmonary penetration of imipenem. *Abstr. 17th Int. Congr. Chemother.* (Berlin).

Moore, R.D., Smith, C.R. and Lietman, P.S. (1984). Association of aminoglycoside plasma levels with therapeutic outcome in Gram-negative pneumonia. *Am. J. Med.* 77, 657–662.

Rouveix, E., Bure, A.M., Regnier, B., Wolff, M., Pangon, B., Laisne, M.J. and Vachon, F. (1986). Experience with imipeneme/cilastatin in the intensive care unit. *J. Antimicrob. Chemother.* 15 (Suppl. E), 153–160.

Schönwald, S., Beus, I., Lisic, M., Car, W. and Gmajnicki, B. (1989). Brief report: ciprofloxacin in the treatment of Gram-negative bacillary meningitis. *Am. J. Med.* 87 (Suppl. 5A), 248–249.

Segev, S., Rosen, N., Joseph, G., Alpern-Elran, H. and Rubinstein, E. (1990). Pefloxacin efficacy in Gram-negative bacillary meningitis. *J. Antimicrob. Chemother.* 26 (Suppl. B), 187–192.

Siegman-Igra, Y., Bar-Yosef, S., Gorea, A. and Avram, J. (1993). Nosocomial *Acinetobacter* meningitis secondary to invasive procedures: report of 25 cases and review. *Clin. Infect. Dis.* 17, 843–849.

Smego, R.A. (1985). Endemic nosocomial *Acinetobacter calcoaceticus* bacteraemia. Clinical significance, treatment, and prognosis. *Arch. Intern. Med.* 145, 2174–2179.

Tilley, P.A.G. and Roberts, F.J. (1994). Bacteraemia with *Acinetobacter* species: risk factors and prognosis in different clinical settings. *Clin. Infect. Dis.* 18, 896–900.

Timsit, J.F., Garrait, V., Misset, B., Goldstein, F.W., Renaud, B. and Carlet, J. (1993). The digestive tract is a major site for *Acinetobacter baumannii* colonization in intensive care patients. *J. Infect. Dis.* 168, 1336–1337.

Urban, C., Go, E., Mariano, N., Berger, B.J., Avraham, I., Rubin, D. and Rahal, J.J. (1993). Effect of sulbactam on infections caused by imipenem-resistant *Acinetobacter calcoaceticus* biotype *anitratus*. *J. Infect. Dis.* 167, 448–451.

Vandenbroucke-Grauls, C.M.J.E. (1993). The threat of multiresistant microorganisms. *Eur. J. Clin. Microbiol. Infect. Dis.* 12 (Suppl. 1), 27–30.

Wolff, M. and Nicolas, M.H. (1994). Les enseignements de l'étude EPIIC. *Lett. Infect.* (Spécial), 15–20.

Appendices

CONTENTS

0-8493-9223-3/95/$0.00+$.50

APPENDIX I

Evolution of Taxonomy and Designations of *Acinetobacter* spp.

	Genus *Moraxella*									
	Group 2 Oxidase-negative						Group 1 Oxidase-positive			
	M. glucidolytica var. non *liquefaciens* var. *liquefaciens*[a]			*M. lwoffii* var. non *liquefaciens* var. *liquefaciens*[a]			*M. lacunata* var. *typica* var. *atypica*		*M. duplex* var. non *liquefaciens* var. *liquefaciens*	
Previous designations	*Bacterium anitratum*, B5W *Diplococcus mucosus*			*Mima polymorpha* *Herellea vaginicola*						
Previous nomenclatures	*Moraxella glucidolytica* genus *Acinetobacter:* one species *Acinetobacter calcoaceticus* (Brisou and Prévôt, 1954)[b]			*Moraxella lwoffii*						
Biovars	*A. anitratus* (glucose-acidifying)						*A. lwoffii*[c] (glucose-negative)			
(Bouvet et al., 1986)	*A. calcoaceticus*	*A. baumannii*	UN[d]	*A. haemolyticus*	*A. junii*	UN[d]	*A. johnsonii*	*A. lwoffii*	UN[d]	*A. radioresistens*
Genomic species[e]	1	2	3	4	5	6	7	8/9	10 11	12

[a] All strains designated *A. liquefaciens* were proteolytic (gelatin hydrolysis which is characteristic of *A. haemolyticus* and genomic species 6 in the current nomenclature).

[b] *Bergey's Manual*, 1984.

[c] *A. alcaligenes* if haemolytic.

[d] UN = unnamed genomic species.

[e] The numbered genomic species are based on DNA-DNA hybridisation results.

APPENDIX II

SELECTIVE ISOLATION OF *Acinetobacter* spp. BY ENRICHMENT CULTIVATION

MEDIUM

Dissolve:

1.5 g KH_2PO_4 in 400 ml distilled water (solution A)
16.5 g $Na_2HPO_4{\cdot}2H_2O$ in 500 ml distilled water (solution B)
5 mg $FeSO_4$ $7H_2O$ in 10 ml distilled water (solution C)

Add to solution A (dissolve each component before adding the next):

0.2 g $MgSO_4{\cdot}7H_2O$
2.0 g NH_4Cl
10 mg $CaCl_2$
1 ml of solution C
2.0 g sodium acetate

Add solution B while stirring. Add distilled water to 960 ml, adjust pH to 7.5, and then adjust final volume to 1000 ml. Autoclave for 15 min at 118°C.

The medium should be tested both for sterility and its nutritional quality with a known *Acinetobacter* test strain.

INOCULATION AND INCUBATION

For swab samples, shake swab briefly in 20–40 ml of the enrichment medium; for liquid samples, add small volumes (e.g., 2 ml) to the medium. Incubate with shaking at 30°C for 24 h. Inoculate a loopful of the enrichment culture onto a solid agar medium (e.g., MacConkey, blood agar or a selective medium) and incubate for a further 24 h at 30°C. Identify presumptive *Acinetobacter* colonies.

REFERENCES

Baumann, P. (1968). Isolation of *Acinetobacter* from soil and water. *J. Bacteriol.* 96, 39–42.

Dijkshoorn, L., van Vianen, W., Degener, J.E. and Michel, M.F. (1987). Typing of *Acinetobacter calcoaceticus* strains isolated from hospital patients by cell envelope protein profiles. *Epidemiol. Infect.* 99, 659–667.

APPENDIX III

Characterisation of Acquired Resistance Mechanisms by Antibiograms

All breakpoints and disc concentrations used in this appendix are those recommended currently by the Comité de l'Antiobiogramme de la Société Française de Microbiologie (Acar et al., 1994).

β–LACTAM RESISTANCE

Low-level production of a cephalosporinase is recognised as follows:

Resistance to amoxycillin, amoxycillin+clavulanic acid, cephalothin, cefoxitin and cefoperazone
Reduced susceptibility to latamoxef and aztreonam
Susceptibility to sulbactam

Production of a penicillinase is recognised as follows:

High-level resistance (inhibition zone of <8 mm for ticarcillin and piperacillin)
Reduced susceptibility to sulbactam
Synergy between ticarcillin and clavulanic acid
Susceptibility to cefotaxime, ceftriaxone, ceftazidime and imipenem

Hyper-production of a cephalosporinase

Reduced susceptibility to ticarcillin
Absence of synergy between ticarcillin and clavulanic acid
Reduced susceptibility to piperacillin, cefotaxime, ceftriaxone and ceftazidime
Susceptibility to sulbactam and imipenem

Reduced permeability

Inhibition zones reduced more for ticarcillin, cefotaxime and ceftazidime than for piperacillin
Association with imipenem resistance and/or non-ß-lactam antibiotics

Combined resistance mechanisms

Resistance to all ß-lactams and/or imipenem (combination of a cephalosporinase, a penicillinase and reduced permeability)
In practice:
Ticarcillin zone size of <15 mm
Cefotaxime/ceftazidime zone size of ≤20 mm
Combination of sulbactam+ticarcillin+clavulanic acid should be tested

AMINOGLYCOSIDES

Many phenotypes can be observed, but certain disc tests are particularly useful: i.e., gentamicin, tobramycin, netilmicin, kanamycin and amikacin+neomycin.

Resistance to neomycin, kanamycin: APH(3')I; if reduced susceptibility to gentamicin, netilmicin: AAC(3)Ia
Resistance to tobramycin, amikacin, netilmicin: AAC(6')I
Resistance to all six aminoglycosides: APH(3')VI + AAC(3)V

In practice:

If gentamicin^S and tobramycin^R — interpretation: amikacin and netilmicin, intermediate or resistant
If netilmicin^R and tobramycin^S — interpretation: gentamicin^R

QUINOLONES

Resistance to pefloxacin and/or ofloxacin
In practice:

If pefloxacin^S, then also ofloxacin^S
If pefloxacin^I (with a zone of >15 mm), then either ofloxacin^S or ^{I}
If pefloxacin^R (with a zone of <15 mm), then also ofloxacin^R

Alternative phenotype shows full resistance to all fluoroquinolones

OTHER ANTIBIOTICS

Acquired resistance to other antibiotics recognised as follows:

Tetracyclines: inhibition zone of 8–25 mm
Trimethoprim and fosfomycin: absence of inhibition zone
Rifampicin: inhibition zone of <22 mm
Sulphonamides: inhibition zone of <28 mm

REFERENCES

Acar, J., Bergogne-Bérézin, E., Chardon, H., Cluzel, R., Courvalin, P., Dabernat, H., Drugeon, H., Dubreuil, L., Duval, J., Flandrois, J.P., Goldstein, F., Meyran, M., Morel, Cl., Philippon, A., Sirot, J., Soussy, C.J. and Thabaut, A. (1994). Comité de l'Antiobiogramme de la Société Française de Microbiologie. Communiqué 1994. *Pathol. Biol.* 42, i–viii.

Bergogne-Bérézin, E. and Joly-Guillou, M.L. (1986). Comparative activity of imipenem, ceftazidime and cefotaxime against *Acinetobacter calcoaceticus*. *J. Antimicrob. Chemother.* 18 (suppl. E), 35–39.

Bergogne-Bérézin, E. and Joly-Guillou, M.L. (1991). Antibiotic resistance mechanisms in *Acinetobacter*. In *The Biology of* Acinetobacter, pp. 83–115. K.J. Towner, E. Bergogne-Bérézin and C.A. Fewson (Eds). Plenum Press, New York.

Bergogne-Bérézin, E., Joly, M.L., Moreau, N. and Le Goffic, F. (1980). Aminoglycoside-modifying enzymes in clinical isolates of *Acinetobacter calcoaceticus*. *Curr. Microbiol.* 4, 361–364.

Buisson, Y., Tran van Nhieu, G., Ginot, L., Bouvet, P., Shill, H., Driot, L. and Meyran, M. (1990). Nosocomial outbreaks due to amikacin-resistant tobramycin-sensitive *Acinetobacter* species : correlation with amikacin usage. *J. Hosp. Infect.* 15, 83–93.

APPENDIX IV

Antibiogram Typing Combined with Cluster Analysis

MEDIA AND REAGENTS

Use solid media that are useful for agar diffusion, e.g., Isosensitest agar (Oxoid CM471)

It is recommended that antibiotic discs or tablets for at least ten different antibiotics should be used. The antibiotics selected should allow discrimination and should not be selected just on the basis of possible therapeutic use.

METHOD

1. Prepare a bacterial suspension of a standardised density (that will subsequently yield confluent or almost confluent growth) in saline from a 24-h culture on blood agar.
2. Inoculate agar plates with the suspension and apply discs or tablets.
3. Incubate for 24 h at 30°C.
4. Measure diameters of the inhibition zones.

CLUSTER ANALYSIS

Use diameters for cluster analysis. By this method the isolates are grouped on the basis of similarity, and this grouping, depicted in a dendrogram, is inspected for correlation with other features, e.g., grouping based on other typing methods, or the time and site of isolation. The Advanced Statistics Facility of the SPSS program can be used for this purpose. Squared Euclidean distance can be used as a (dis)similarity measure, and grouping can be obtained by the UPGMA clustering criterion.

STANDARDISATION AND QUALITY CONTROL

Standardisation of inoculum, incubation time and culture conditions are important for longitudinal reproducibility. One, or more than one, 'in-house' reference strains should be included. These should cluster together at a relatively high similarity level.

REFERENCES

Dijkshoorn, L., Aucken, H.M., Gerner-Smidt, P., Kaufmann, M.E., Ursing, J. and Pitt, T.L. (1993). Correlation of typing methods for *Acinetobacter* isolates from hospital outbreaks. *J. Clin. Microbiol.* 31, 702–705.

Norusis, M.J. (1988). *SPSS/PC+ Advanced Statistics™ V2.0, B-71 – B-101*. SPSS Inc., Chicago.

Sneath, P.H.A. and Sokal, R.R. (1973). *Numerical Taxonomy*. W.H. Freeman, San Francisco.

APPENDIX V

Protein SDS-PAGE

CELL ENVELOPE PREPARATION

Media and reagents:

Nutrient broth No. 2 (Oxoid CM67)
Sonication buffer: 50 m*M* Tris-HCl, 2 m*M* EDTA, pH 8.5
Pellet buffer: 2 m*M* Tris-HCl, pH 7.8
Protein assay kit 500-0006 (Bio-Rad)

Sample mix reagent (per cell envelope protein sample):

20 µl 0.625 *M* Tris-HCl, pH 6.8
40 µl SDS 10% w/v
40 µl Glycerol 50% w/v
20 µl Bromophenol blue 0.01% w/v
10 µl 2-Mercaptoethanol

Method:

1. Culture cells in 40 ml nutrient broth no. 2 at 30°C for 20 h with aeration.
2. Harvest cells by centrifugation for 10 min at 7000 g.
3. Resuspend cells in 4 ml of sonication buffer.
4. Disrupt cells by ultrasonication (3 × 20 sec bursts at an amplitude of 20 µm) at 4°C.
5. Centrifuge for 20 min at 2700 g to remove large fragments and unbroken cells.
6. Collect cell envelope fractions from the supernatant by centrifugation for 60 min at 12,300 g.
7. Resuspend the pellet in 100–200 µl of pellet buffer.
8. Store samples at -20°C or use directly.

Protein determination: determine protein concentration of cell envelope fractions by means of the protein assay kit, with bovine albumin used for calibration purposes.

Sample preparation for SDS-PAGE:

1. Take cell envelope sample containing 100 μg protein and add distilled water to 70 μl (final volume).
2. Add 130 μl of sample mix reagent (to give a final composition of 0.0625 *M* Tris-HCl, pH 6.8, SDS 2%, glycerol 10%, bromophenol blue 0.001%, 2-mercaptoethanol 5% and protein 0.5 μg/ml.
3. Boil samples for 5 min.
4. For repeated use, store samples at –20°C and boil again before application to the gel (see below).

Quality control: one 'in-house' reference strain should be included in each batch of cell envelope preparations, and its eventual electrophoretic pattern compared with previous patterns obtained with the same strain.

WHOLE-CELL PROTEIN PREPARATION

Media and reagents:

Tryptone soya agar (TSA; Becton-Dickinson) plates
Phosphate-buffered saline (NaPBS)
Treatment solution: Tris, 0.75 g; 2-mercaptoethanol, 5 ml; glycerol, 10 ml; 0.01% bromophenol blue, 10 ml; distilled water to 100 ml; pH 6.8

Whole-cell preparation:

1. Culture cells on TSA for 24 h at 30°C.
2. Suspend cells in 30 ml NaPBS.
3. Centrifuge cells for 20 min at 7000 g and 4°C.
4. Wash cells with 30 ml NaPBS and recentrifuge for 10 min at 7000 g and 4°C.
5. Resuspend 100 mg of wet cells in 1.8 ml of treatment solution (see above) and vortex to mix.
6. Add 0.2 ml SDS 20% and vortex to mix.
7. Heat at 95°C for 10 min, and then centrifuge for 5 min at 7000 g and 4°C.
8. Store supernatant at –20°C or use directly.

PREPARATION OF CALIBRATION PROTEINS

Standard proteins (Sigma):

Phosphorylase B (97 kDa)
Bovin albumin (66 kDa)
Ovalbumin (45 kDa)
Glyceraldehyde-3-phosphate dehydrogenase subunit (36 kDa)
Carbonic anhydrase (29 kDa)
Trypsin inhibitor (20 kDa)
Lactalbumin (14 kDa)

Method:

1. Prepare a 700-μl mixture containing 100 μg of each standard protein and add 1300 μl of sample mix (see Cell Envelope Preparation).
2. Heat samples according to the manufacturer's instructions and store small volumes at –20°C.

POLYACRYLAMIDE GEL ELECTROPHORESIS

Reagents:

Stock sol. I: 44 g acrylamide + 0.8 g methylene *bis*acrylamide dissolved in 100 ml distilled H_2O
Stock sol. II: 30 g acrylamide + 0.8 g methylene *bis*acrylamide dissolved in 100 ml distilled H_2O
Separation gel buffer: 0.75 *M* Tris-HCl, pH 8.8 (buffer I)
Stacking gel buffer: 0.25 *M* Tris-HCl, pH 6.8 (buffer II)
Ammonium persulphate 10 mg/ml
SDS 10% w/v
Isobutanol
Electrode buffer: 0.025 *M* Tris, 0.19 *M* glycine, 0.1% SDS, pH 8.3

Equipment:

Protean II 16 × 16 cm slab gel equipment (Bio-Rad) with 1.5-mm thick gels

Electrophoresis:

1. Prepare running gel containing 10 ml stock sol. I; 8.2 ml distilled H_2O; 20 ml buffer I; 750 μl ammonium persulphate 10 mg/ml.
2. De-aerate and add 800 μl SDS 10%; 60 μl *N,N,N',N'*-tetramethylenediamine (TMED).

3. Pour gel and immediately apply 200 µl isobutanol to surface.
4. After 1 h, replace isobutanol with 600 µl of running gel buffer reagent without stock sol. I (as above) and cover with parafilm to prevent evaporation.
5. Allow to polymerise for a standard period of at least 16 h at room temperature.
6. Prepare running gel containing 1 ml stock sol. II; 3.65 ml distilled H_2O; 5 ml buffer II; 250 µl ammonium persulphate 10 mg/ml. De-aerate and add 100 µl SDS 10%; 10 µl *N,N,N′,N′*-tetramethylenediamine (TMED).
7. Pour separation gel with comb (adjusted as necessary).
8. Allow to polymerise for a standard period of at least 1 h at room temperature.
9. Apply 35 µl of sample to each slot.

Quality control:

Include standard molecular weight markers and at least one standard 'in-house' reference strain with a constant and wide-ranging banding pattern on each gel.

REFERENCES

Dijkshoorn, L., Michel, M.F. and Degener, J.E. (1987). Cell envelope protein profiles of *Acinetobacter calcoaceticus* strains isolated in hospitals. *J. Med. Microbiol.* 23, 313–319.

Dijkshoorn, L., Tjernberg, I., Pot, B., Michel, M.F., Ursing, J. and Kersters, K. (1990). Numerical analysis of cell envelope protein profiles of *Acinetobacter* strains classified by DNA-DNA hybridization. *Syst. Appl. Microbiol.* 13, 338–344.

Kersters, K. and De Ley, J. (1975). Identification and grouping of bacteria by numerical analysis of their electrophoretic protein patterns. *J. Gen. Microbiol.* 87, 333–342.

Lugtenberg, B., Meijers, J., Peters, R., van der Hoek, P. and van Alphen, L. (1975). Electrophoretic resolution of the 'major outer membrane protein' of *Escherichia coli* K12 into four bands. *FEBS Lett.* 58, 254–258.

APPENDIX VI

Pulsed-Field Gel Electrophoresis (PFGE)

MEDIA, REAGENTS AND EQUIPMENT

Nutrient broth no. 2 (Oxoid CM67), supplemented with $NaNO_3$ 0.1% w/v
SE buffer: 75 m*M* NaCl; 25 m*M* EDTA; pH 7.5
TE buffer: 10 m*M* Tris; 1 m*M* EDTA; pH 8.0
Lysis buffer: *N*-lauroylsarcosine 1% w/v; 0.5 *M* EDTA; pH 9.5
Proteinase K 50 mg/ml; stored at -20°C in 200-µl volumes
Low melting temperature (LMT) agarose
Restriction enzymes (and reaction buffers) suitable for *Acinetobacter* spp. (e.g., *Apa*I, *Sma*I, *Sfi*I or *Nhe*I)
Bovine serum albumin (molecular biology grade) 10 mg/ml
10× TBE buffer stock solution (0.89 *M* Tris; 0.89 *M* boric acid; 20 m*M* EDTA); dilute 1:20 for use
Agarose (molecular biology grade)
Ethidium bromide 1 mg/ml

Equipment: e.g., CHEF-DRII apparatus (Bio-Rad) with accessories.

PREPARATION OF DNA

1. Mix equal volumes of a bacterial suspension (5×10^9 cfu/ml) and LMT agarose solution (2% in SE buffer) in a 50°C waterbath.
2. Allow the mixture to set at 4°C in the slots of the CHEF mould.
3. Transfer the plugs to a bijou bottle and add 2 ml of lysis buffer, followed by proteinase K to a final concentration of 500 µg/ml.

4. Incubate overnight with gentle shaking at 56°C.
5. Replace with fresh buffer and proteinase K; repeat incubation step.
6. Discard lysis buffer, add TE buffer and keep at 4°C for 30 min; repeat this step twice more.
7. Plugs can then be kept in TE buffer at 4°C for several months.

RESTRICTION ENZYME DIGESTION

1. Cut small portions (3 mm × 4.5 mm) of plugs and place in the wells of a microtitre plate or in 1.5-ml microtitre tubes.
2. Add 100 µl of enzyme reaction buffer and leave at 4°C for 30 min.
3. Replace with fresh reaction buffer, BSA 0.1 mg/ml and 5 U of an appropriate restriction enzyme.
4. Mix gently and incubate according to the manufacturer's instructions.
5. Store the digests in TE buffer at 4°C until required.

ELECTROPHORESIS

1. Prepare an agarose 1.2% w/v gel in 0.5× TBE buffer.
2. Place the digested plugs against the leading edge of the wells with a scalpel or spatula; seal the wells and plugs with molten agarose (60°C).
3. Place the gel in the electrophoresis chamber and add 0.5× TBE buffer to just cover the gel; check that no bubbles remain underneath the gel.
4. Perform electrophoresis; conditions of 200 V for 24 h at 16°C have been found to be useful, but should be varied according to the sizes of the fragments to be separated. Initial pulse time, 1 sec; final pulse time, 35 sec.

STAINING OF PFGE GELS

1. Stain the gel in ethidium bromide (1 µg/ml) for 1 h.
2. Destain in distilled water for 12–24 h at 4°C.
3. Photograph with UV transillumination.

REFERENCES

Allardet-Servent, A., Bouziges, N., Carles-Nurit, M.J., Bourg, G., Gouby, A. and Ramuz, M. (1989). Use of low-frequency cleavage restriction endonucleases for DNA analysis in epidemiological investigations of nosocomial bacterial infections. *J. Clin. Microbiol.* 27, 2057–2061.

Birren, B. and Lai, E. (1993). *Pulsed Field Gel Electrophoresis*. Academic Press, London.

Gouby, A., Carles-Nurit, M.J., Bouziges, N., Bourg, G., Mesnard, R. and Bouvet, P.J.M. (1992). Use of pulsed-field gel electrophoresis for investigation of hospital outbreaks of *Acinetobacter baumannii*. *J. Clin. Microbiol.* 30, 1588–1591.

Gräser, Y., Klare, I., Halle, E., Gantenberg, R., Buchholz, P., Jacobi, H.D., Presber, W. and Schönian, G. (1993). Epidemiological study of an *Acinetobacter baumannii* outbreak by using polymerase chain reaction fingerprinting. *J. Clin. Microbiol.* 31, 2417–2420.

Struelens, M.J., Carlier, E., Maes, N., Serruys, E., Quint, W.G.V. and van Belkum, A. (1993). Nosocomial colonization and infection with multiresistant *Acinetobacter baumannii*: outbreak delineation using DNA macrorestriction analysis and PCR-fingerprinting. *J. Hosp. Infect.* 25, 15–32.

APPENDIX VII

Polymerase Chain Reaction (PCR) Fingerprinting

MEDIA, REAGENTS AND EQUIPMENT

TSB agar (Oxoid CM131) plates
Sterile phosphate-buffered saline, pH 7.4 (PBS)
TEG buffer: 50 m*M* glucose; 10 m*M* EDTA; 25 m*M* Tris-HCl; pH 8.0
SDS 10% w/v
Lysozyme 125 mg/ml
Proteinase K stock solution (10 mg/ml). Store at –20°C; dilute 10-fold before use
Phenol
Phenol:chloroform:isoamyl alcohol (25:24:1)
Chloroform:isoamyl alcohol (24:1)
3 *M* sodium acetate, pH 5.4
Ethanol 96% v/v
Ethanol 70% v/v
PCR reaction mixture (per 100-µl reaction): 10 m*M* Tris-HCl, pH 9.0; 50 m*M* KCl; 2.5 m*M* $MgCl_2$; gelatin 0.01%; Triton X-100 .1%; dNTPs at a final concentration of 0.2 m*M* each; 0.625 U *Taq* DNA polymerase; primers (see below) at a concentration of 100 pmol per reaction (or 50 pmol per reaction if used in combination)
Primers: primers ERIC1 and ERIC2 (Versalovic et al., 1991) have been found to be useful for fingerprinting *Acinetobacter* strains (Struelens et al., 1993; L. Dijkshoorn and B. Giesendorf,

unpublished results). Alternative primers have been used by Gräser et al. (1993)

Equipment: thermocycler; agarose gel electrophoresis equipment

EXTRACTION OF NUCLEIC ACIDS

1. Inoculate a TSB agar plate with a fresh overnight culture and incubate overnight at 37°C.
2. Harvest cells and resuspend in 2 ml PBS.
3. Mix 100 µl of cell suspension with 150 µl TEG buffer.
4. Incubate for 5 min at room temperature with 10 µl lysozyme solution (125 mg/ml).
5. Add 12.5 µl proteinase K solution (1 mg/ml) and 12.5 µl of SDS 10% w/v; incubate for 30 min at 37°C.
6. Extract nucleic acids with phenol, followed by phenol:chloroform: isoamyl alcohol (25:24:1), followed by chloroform:isoamyl alcohol (24:1).
7. Precipitate nucleic acids with 0.1 vol of 3 *M* sodium acetate and 2.5 vol ethanol 96% v/v.
8. Wash pellet with 70% v/v ethanol, dry and resuspend in double-distilled water.
9. Estimate nucleic acid concentration on an agarose 1% w/v gel with known amounts of bacteriophage lambda DNA as a reference standard.
10. Adjust to a DNA concentration of 10 ng/µl; use 10 µl per amplification reaction.

PCR REACTIONS

1. Add 100 ng of DNA sample to each reaction mixture.
2. Incubate for 2 min at 94°C before amplification; amplify with 35 cycles of 1 min @ 94°C, 1 min @ 25°C, and 4 min @ 74°C, followed by a final extension period of 10 min @ 74°C.
3. Analyse amplification products on agarose 1.2% w/v gels, with ethidium bromide staining followed by UV transillumination.

Quality control: include one 'in-house' reference strain in each batch of nucleic acid extractions and PCR reactions. Distilled water should be used as a negative control in the amplification reactions. A

1-kb DNA ladder (e.g., Gibco/BRL) should be used to estimate fragment sizes on agarose gels.

REFERENCES

Giesendorf, B.A.J., Goossens, H., Niesters, H.G.M., van Belkum, A., Koeken, A., Endtz, H.P., Stegeman, H. and Quint, W.G.V. (1994). Polymerase chain reaction-mediated DNA fingerprinting for epidemiological studies on *Campylobacter* spp. *J. Med. Microbiol.* 40, 141–147.

Gräser, Y., Klare, I., Halle, E., Gantenberg, R., Buchholz, P., Jacobi, H.D., Presber, W. and Schönian, G. (1993). Epidemiological study of an *Acinetobacter baumannii* outbreak by using polymerase chain reaction fingerprinting. *J. Clin. Microbiol.* 31, 2417–2420.

Struelens, M.J., Carlier, E., Maes, N., Serruys, E., Quint, W.G.V. and van Belkum, A. (1993). Nosocomial colonization and infection with multiresistant *Acinetobacter baumannii*: outbreak delineation using DNA macrorestriction analysis and PCR-fingerprinting. *J. Hosp. Infect.* 25, 15–32.

Versalovic, J., Koeuth, T. and Lupski, J.R. (1991). Distribution of repeditive DNA sequences in eubacteria and application to fingerprinting of bacterial genomes. *Nucl. Acids Res.* 19, 6823–6831.

APPENDIX VIII

RIBOTYPING

Contributed by P. Gerner-Smidt

REAGENTS AND EQUIPMENT

Lysis buffer (50 m*M* Tris, 50 m*M* EDTA, pH 8.0)
Lysozyme
Proteinase K
SDS
Phenol:chloroform:isoamyl alcohol (25:24:1)
Isopropanol
TE buffer (10 m*M* Tris, 1 m*M* EDTA, pH 8.0)
Ammonium acetate
Ethanol 96% v/v
Restriction enzymes
Colour mix (sucrose 50% w/v, bromophenol blue 0.25% w/v in 0.02 *M* Tris, 0.02 *M* EDTA, pH 8.0)
Agarose
Ethidium bromide
TBE buffer (89 m*M* Tris, 89 m*M* boric acid, 2 m*M* EDTA, pH 8.0)
HCl
NaOH
Positively charged nylon membranes (e.g., Hybond N+, Amersham)
2× SSC buffer (0.3 *M* NaCl, 0.03 *M* sodium citrate, pH 7.0)
16S and 23S rRNA from *E. coli* (Boehringer Mannheim)
M-MLV reverse transcriptase (Gibco BRL)
DIG labelling and detection reagents (Boehringer Mannheim)

Electrophoresis equipment: vacuum or pressure blotting equipment

DNA PURIFICATION

(Mekalanos, 1983)

1. Resuspend the pellet from an overnight broth culture in 500 µl lysis buffer; add 20 µl lysozyme (10 mg/ml) and incubate at 37°C for 30 min.
2. Lyse the cells by gentle mixing of the suspension with 50 µl proteinase K (10 mg/ml) and 20 µl SDS 10% w/v at 56°C until the lysate clears (ca. 30–120 min).
3. Remove protein by extracting with 1 vol of phenol:chloroform: isoamyl alcohol (25:24:1); mix well and then centrifuge in a microfuge tube at 10,000 g for 10 min; transfer the upper aqueous phase to a fresh tube (with a plastic pipette tip cut to make the tip aperture bigger); repeat this entire procedure once.
4. Precipitate the DNA by the addition of 1 vol isopropanol. Leave at room temperature for 5 min and then centrifuge for 5 min at 10,000 g; discard the supernatant and dry the pellet.
5. Redissolve the DNA in 300 µl TE buffer and allow to stand at room temperature for 5 min; then reprecipitate the DNA by the addition of 100 µl 7.5 *M* ammonium acetate and 800 µl ethanol 96% v/v; place the mixture at 4°C for at least 30 min (may be left overnight).
6. Recover the DNA by centrifugation at 10,000 g for 5 min and discard the supernatant; estimate the amount of DNA from the size of the pellet.
7. Dry the pellet well and then redissolve in an appropriate amount of TE buffer (usually between 50 and 500 µl); check the final DNA concentration in a spectrophotometer.

RESTRICTION ENDONUCLEASE DIGESTION

1. A range of enzymes can be used (see Gerner-Smidt, 1992); digest an appropriate amount of DNA (1–3 µg) following the manufacturer's instructions.
2. Add 2–3 µl colour mix to each digestion before electrophoresis.

ELECTROPHORESIS

1. Prepare agarose in TBE buffer; for fragments generated by enzymes with a 5 or 6 base recognition sequence, a 0.7–0.8% gel is used; for enzymes with a 4 base recognition sequence, a 1% gel is used.

2. Once set, cover gel with TBE buffer and then load 10–20 µl of DNA sample per track.
3. Perform electrophoresis overnight at 50 V (2–2.5 V/cm).
4. Stain gel with ethidium bromide (2 mg/L) for 30 min; examine DNA bands by UV transillumination to check completeness of digestion and intensity of bands, etc.; if all is satisfactory, proceed to blotting step.

BLOTTING OF DNA FRAGMENTS TO MEMBRANE

1. Place the gel in a plastic tray and cover with 0.25 *M* HCl; shake gently for 15 min, then wash twice with distilled water.
2. Follow the manufacturer's instructions for using the blotting equipment to transfer DNA fragments to a positively charged nylon membrane (e.g., Hybond-N+); use an alkaline blotting solution (0.5 *M* NaOH, 0.6 *M* NaCl).
3. Wash the membrane twice in 2× SSC buffer and then allow to dry.

PROBE PREPARATION

(May be scaled up ×10)

1. Prepare a mixture of 1.5 µg 16S and 23S RNA from *E. coli* (Boehringer Mannheim), 0.1 m*M* dATP, 0.1 m*M* dCTP, 0.1 m*M* dGTP, 0.064 m*M* dTPP, 0.036 m*M* DIG-11-dUTP (Boehringer Mannheim), 0.5 m*M* random hexanucleotides, 200 U reverse transcriptase (Gibco), 10 m*M* dithiothreitol, 3 m*M* $MgCl_2$ and 50 m*M* Tris-HCl, pH 8.3, in a total volume of 20 µl. Incubate for 2 h–overnight at 37°C.
2. Stop the reaction by adding 2 µl 0.2 *M* EDTA (pH 8.0); purify the probe by means of the procedures detailed in the *DIG DNA Labelling and Detection Applications Manual* (Boehringer Mannheim).

PREHYBRIDISATION, HYBRIDISATION AND DETECTION

1. The procedures used are essentially those outlined in the *DIG Applications Manual* (Boehringer Mannheim), with hybridisation being performed overnight at 60°C.
2. Fragments hybridising with the labelled probe are detected by a colorimetric reaction, as described in the *DIG Applications Manual*,

usually within 2–3 h, but the membrane can also be left to develop overnight without increasing the background staining significantly.

3. The colour reaction is terminated by washing the membrane in tap water.

REFERENCES

Gerner-Smidt, P. (1992). Ribotyping of the *Acinetobacter calcoaceticus - Acinetobacter baumannii* complex. *J. Clin. Microbiol.* 30, 2680–2685.

Mekalanos, J.J. (1983). Duplication and amplification of toxin genes in *Vibrio cholerae*. *Cell* 35, 253–263.

APPENDIX IX

Plasmid Isolation Procedure

Contributed by P. Gerner-Smidt

REAGENTS

STET buffer (sucrose 8% w/v, Triton® X-100 0.5% v/v, 50 m*M* EDTA, 10 m*M* Tris, pH 8.0)
Lysozyme
Chloroform
3 *M* ammonium acetate
Isopropanol
TBE buffer (89 m*M* Tris, 89 m*M* boric acid, 2 m*M* EDTA, pH 8.0)
Colour mix (sucrose 50% w/v, bromophenol blue 0.25% w/v, in 0.02 *M* Tris, 0.02 *M* EDTA, pH 8.0)
Ethidium bromide

METHOD

1. Suspend half a 10-µl loopful of growth from an overnight plate culture in 350 µl of STET buffer in a microfuge tube.
2. Add 25 µl of lysozyme solution (400 mg in 1 ml STET buffer) and incubate at 37°C until the suspension clears and becomes viscous (ca. 1 min).
3. Boil the suspension in a water bath for 40 s and then centrifuge at 10,000 g for 10 min at 4°C.
4. Discard the pellet with a toothpick; extract protein from the supernatant by adding 350 µl chloroform, mixing, and centrifuging at 10,000 g for 4 min at 4°C.
5. Transfer the upper aqueous layer to a fresh microfuge tube; precipitate the DNA by adding 35 µl 3 *M* ammonium acetate, 350 µl

isopropanol, and mixing by inversion; leave at -20°C overnight or at -80°C for 2 h.

6. Recover the DNA by centrifuging at 10,000 g for 20 min at 4°C; discard the supernatant and dry the pellet.
7. Dissolve the pellet in 20 µl TBE buffer and add 2 µl colour mix.
8. Electrophorese overnight at 50 V (2–2.5 V/cm) on an agarose 0.8% w/v gel; stain the gel with ethidium bromide (2 mg/L) for 30 min; visualise the plasmid profiles on a UV transilluminator.

REFERENCES

Gerner-Smidt, P. (1989). Frequency of plasmids in strains of *Acinetobacter calcoaceticus*. *J. Hosp. Infect.* 14, 23–28.

INDEX

B

C

D

Q

R